普通高等教育"十三五"规划教材

新工科工程英语系列教材　总主编　孙昶临

工程英语学术写作基础

刘纪红　张聪颖　主编

北　京

冶金工业出版社

2020

内 容 提 要

本教材旨在帮助理工院校本科生（尤其是工程类各专业）了解英语学术论文写作的整体结构，为以后的学术论文写作打下基础。教材的主体内容分为 7 个章节，依次为总论、标题和关键词、摘要、文献综述、论文主体写作、参考文献引注规范和学术论文写作中的问题汇总。其特色在于，教材内大量的英文文献阅读解析，一方面贴合专业内容，让学术论文不再讳莫如深，另一方面也提升了理科生的人文素养。另外，为了拓宽学生的阅读范围，也为了实际写作需要，本教材把中文科技论文参考文献的运用及其著录规则作为附录附在书后。

本教材可作为工程类院校英语课程的主干教材，也可服务于工程科研领域的广大作者，还可为所有工程类英语爱好者提供高质量的学习素材。

图书在版编目(CIP)数据

工程英语学术写作基础/刘纪红，张聪颖主编．—北京：冶金工业出版社，2020.6

普通高等教育"十三五"规划教材
ISBN 978-7-5024-8497-2

Ⅰ.①工⋯ Ⅱ.①刘⋯ ②张⋯ Ⅲ.①工程技术—英语—论文—写作—高等学校—教材 Ⅳ.①TB

中国版本图书馆 CIP 数据核字(2020)第 086025 号

出 版 人　陈玉千
地　　址　北京市东城区嵩祝院北巷 39 号　邮编　100009　电话　(010)64027926
网　　址　www.cnmip.com.cn　电子信箱　yjcbs@cnmip.com.cn
责任编辑　夏小雪　美术编辑　吕欣童　版式设计　孙跃红
责任校对　郑　娟　责任印制　李玉山

ISBN 978-7-5024-8497-2

冶金工业出版社出版发行；各地新华书店经销；三河市双峰印刷装订有限公司印刷
2020 年 6 月第 1 版，2020 年 6 月第 1 次印刷
787mm×1092mm　1/16；9.25 印张；222 千字；137 页
40.00 元

冶金工业出版社　　投稿电话　(010)64027932　投稿信箱　tougao@cnmip.com.cn
冶金工业出版社营销中心　电话　(010)64044283　传真　(010)64027893
冶金工业出版社天猫旗舰店　yjgycbs.tmall.com

(本书如有印装质量问题，本社营销中心负责退换)

本书编写人员
（排名不分先后顺序）

总主编 孙昶临

主　编 刘纪红　张聪颖

副主编 程　舒　邢春燕　王振华　邵明霞　冯宗侠

编　委 杨　晶　郭红梅　崔　磊　迟丽君　马红敬

　　　　　刘海燕　赵　蔚　胡　鹏　杨瑞芝　韩　芳

前　言

新工科对课程分类与设置提出了新要求，学科交叉融合成为体现这一要求的关键。我国大学英语教育在全球化和信息化背景下肩负培养高素质跨文化交际人才的重任，在大学英语教学改革的进程中，突出"专业+英语"，提高学生英语应用能力，成为大学英语教学的重中之重。

《工程英语学术写作基础》以教育部颁布的《非英语专业英语教学大纲》为依据，以大学英语课程教学要求为宗旨，突出工程类英语的鲜明特色，以英语为载体，传递工程领域的研究成果及研究走向，汇集了大量的工程类词汇与表达方式。在强调英语综合技能训练的同时，更注重较高阶段英语应用能力的培养，帮助学生了解英语国家的学术规范和学术英语写作的基本要求，有助于学习者顺利完成在专业课学习中所涉及的论文查写、案例分析、研究报告等方面的学术英语写作任务，运用规范的语言表达自己的学术思想。对于工程类毕业生更好地适应我国社会经济的迅猛发展，增强其国际竞争能力，具有重要意义。

教材内容共分七章及附录，主要就学术英语写作的语言要求、引用规范、句子特点、段落组织、篇章构建等方面进行介绍，各章内容既相互联系又相对独立。第一章为总论，第二章为标题和关键词，第三章为摘要，第四章为文献综述，第五章为论文主体写作，第六章为参考文献引注规范，第七章为学术论文写作中的问题汇总。另外，为了拓宽学生的阅读范围，也为了实际写作需要，本教材把中文科技论文参考文献的运用及其著录规则作为附录附在书后。

本书的特色主要体现在两个方面：首先，全书突出工程英语的使用，强调国际学术规范以及学术语境下的英语运用能力，为学习者使用规范的英语交流学术思想奠定基础。所用范例均出自国内外正规期刊文章，为学生提供真实写作样板，有助于学生快速掌握学术论文的基本写作要领及规范、有效阅读高水平学术论文，并通过仿写和创造性写作，培养学生的学术英语写作能力。其次，建立"以学生为中心的互动型、实践型、自主学习"的教学模式，突出培养学生的语言应用能力。强调学术写作过程中的操作与实践，书中设计了大量的课内外练习，让学习者通过语言的运用，不断提高自己的语言能力。

　　本书的编者均为山东建筑大学外国语学院的教师，长期从事大学英语教学工作，特别是在学术英语教学方面进行了积极的探索。全书由孙昶临组织编写，由刘纪红、张聪颖、王振华、邢春燕负责统稿。其中，第一章由张聪颖负责编写；第二章由孙昶临负责编写；第三章由王振华负责编写；第四章由邢春燕负责编写；第五章由程舒负责编写；第六章由刘纪红负责编写；第七章由邵明霞负责编写；书后附录由冯宗侠负责编写。

　　本教材不仅可作为工程类院校英语课程的主干教材，也可服务于工程科研领域的广大工作者，还可为所有工程类英语爱好者提供高质量的学习素材，具有较强的针对性和实用性。

　　我们希望能与广大师生交流相关教学经验和问题。当然，由于编者的水平所限，书中难免存在诸多不足之处，欢迎广大读者和同行批评指正！此外，本书的编写获得了山东建筑大学外国语学院领导和同事们的大力支持，在此表示衷心感谢！

编　者
2020 年 4 月

目 录

1 总论 ·· 1
  1.1 学术论文英语写作的特征 ·· 1
    1.1.1 创新性 ·· 1
    1.1.2 逻辑性 ·· 1
    1.1.3 启发性 ·· 1
    1.1.4 连贯性 ·· 1
    1.1.5 准确性 ·· 2
    1.1.6 规范性 ·· 2
  1.2 学术论文英语写作的语言特征 ··· 2
    1.2.1 语言规范 ··· 2
    1.2.2 句子结构严谨 ·· 2
    1.2.3 大量使用专业词汇 ··· 4
    1.2.4 学术英语的时态 ·· 4
    1.2.5 学术英语的语态 ·· 6
    1.2.6 学术英语的人称 ·· 6
    1.2.7 口语体和书面体 ·· 7
  1.3 论文的组成部分 ·· 8
    1.3.1 标题和摘要 ··· 9
    1.3.2 正文的组成和写作 ··· 10
    1.3.3 致谢和参考文献 ·· 11
  1.4 论文实例 ·· 13
  练习题 ·· 18
  参考文献 ·· 20

2 标题和关键词 ··· 21
  2.1 概述 ··· 21
    2.1.1 标题的定义 ··· 21
    2.1.2 标题的作用 ··· 21
    2.1.3 标题的格式 ··· 21
    2.1.4 关键词的定义 ·· 21
    2.1.5 关键词的作用 ·· 22
    2.1.6 关键词的格式 ·· 22
  2.2 基本要求 ·· 22

- 2.2.1 标题的简洁性要求 ………………………………………………… 22
- 2.2.2 标题的句法要求 …………………………………………………… 23
- 2.2.3 标题的指示性要求 ………………………………………………… 24
- 2.2.4 标题的朴实性要求 ………………………………………………… 24
- 2.2.5 标题中尽量使用关键词 …………………………………………… 24
- 2.2.6 关键词的选取要求 ………………………………………………… 25

2.3 写作范式 …………………………………………………………………… 25
- 2.3.1 标题的构成 ………………………………………………………… 25
- 2.3.2 标题的句法规则 …………………………………………………… 25
- 2.3.3 标题大小写错误 …………………………………………………… 26
- 2.3.4 标题常见语法错误 ………………………………………………… 27
- 2.3.5 关键词的组成类型 ………………………………………………… 28
- 2.3.6 关键词的特殊选择 ………………………………………………… 28

2.4 范例阅读与分析 …………………………………………………………… 29

练习题 …………………………………………………………………………… 30

参考文献 ………………………………………………………………………… 35

3 摘要

3.1 概述 ………………………………………………………………………… 37
- 3.1.1 摘要的定义 ………………………………………………………… 37
- 3.1.2 摘要的分类 ………………………………………………………… 37
- 3.1.3 摘要的功能 ………………………………………………………… 38
- 3.1.4 摘要的基本要素 …………………………………………………… 38

3.2 摘要的写作范式 …………………………………………………………… 39
- 3.2.1 研究主题的写作范式 ……………………………………………… 39
- 3.2.2 研究背景的写作范式 ……………………………………………… 39
- 3.2.3 研究目的的写作范式 ……………………………………………… 40
- 3.2.4 研究方法的写作范式 ……………………………………………… 41
- 3.2.5 研究结果的写作范式 ……………………………………………… 42
- 3.2.6 研究结论的写作范式 ……………………………………………… 43

3.3 范文阅读与分析 …………………………………………………………… 43
- 3.3.1 相关阅读：英文摘要的注意事项 ………………………………… 43
- 3.3.2 范文分析 …………………………………………………………… 46

练习题 …………………………………………………………………………… 51

参考文献 ………………………………………………………………………… 51

4 文献综述

4.1 概述 ………………………………………………………………………… 53
- 4.1.1 文献综述的简介 …………………………………………………… 53

    4.1.2　文献综述的研究意义和目的 …………………………………………… 54
    4.1.3　文献综述的写作 …………………………………………………………… 54
  4.2　文献评估与利用 ……………………………………………………………………… 55
    4.2.1　文献评估 …………………………………………………………………… 55
    4.2.2　文献利用 …………………………………………………………………… 55
  4.3　文献回顾与写作 ……………………………………………………………………… 56
    4.3.1　文献回顾 …………………………………………………………………… 56
    4.3.2　文献写作 …………………………………………………………………… 58
  4.4　范文阅读与分析 ……………………………………………………………………… 60
    4.4.1　文献综述范文导读 ………………………………………………………… 60
    4.4.2　文章中文献综述部分的写法导读 ………………………………………… 70
  练习题 ………………………………………………………………………………………… 72
  参考文献 ……………………………………………………………………………………… 72

5　论文主体写作 …………………………………………………………………………… 73

  5.1　引言 …………………………………………………………………………………… 73
    5.1.1　引言的常见写法 …………………………………………………………… 75
    5.1.2　引言中常见的问题 ………………………………………………………… 75
    5.1.3　引言实例 …………………………………………………………………… 76
  5.2　正文 …………………………………………………………………………………… 77
    5.2.1　正文内容的顺序排列 ……………………………………………………… 78
    5.2.2　主题句的写作方法 ………………………………………………………… 79
    5.2.3　拓展句的写作方法 ………………………………………………………… 80
  5.3　结果和讨论 …………………………………………………………………………… 81
    5.3.1　如何写结果和讨论 ………………………………………………………… 81
    5.3.2　结果和讨论应注意的问题 ………………………………………………… 82
    5.3.3　结果撰写常用句型 ………………………………………………………… 82
    5.3.4　讨论撰写常用句型 ………………………………………………………… 82
  5.4　结论 …………………………………………………………………………………… 83
    5.4.1　结论涵盖的内容 …………………………………………………………… 83
    5.4.2　结论撰写应注意的问题 …………………………………………………… 83
    5.4.3　结论部分的常用句型 ……………………………………………………… 83
  5.5　句子和段落写作中的语言问题 ……………………………………………………… 84
    5.5.1　词汇使用不当 ……………………………………………………………… 84
    5.5.2　句式使用不当 ……………………………………………………………… 85
  练习题 ………………………………………………………………………………………… 86
  参考文献 ……………………………………………………………………………………… 87

6 参考文献引注规范 · · · · · · 88

6.1 概述 · · · · · · 88
6.2 英语学术论文参考文献引注规范简介及部分工程类英语期刊引注格式举例 · · · · · · 89
6.2.1 MLA 与 APA 文内引用常用格式对比 · · · · · · 89
6.2.2 MLA 与 APA 文后参考文献常见著录格式对比 · · · · · · 93
6.2.3 文后参考文献的排列 · · · · · · 95
6.2.4 部分工程类英语期刊引用格式举例 · · · · · · 97
6.3 引用过程中的预防剽窃 · · · · · · 103
6.3.1 什么是剽窃 · · · · · · 103
6.3.2 如何在引用过程中预防剽窃 · · · · · · 104
6.4 范文阅读与分析 · · · · · · 105
6.4.1 范文阅读 · · · · · · 105
6.4.2 范文分析 · · · · · · 106
练习题 · · · · · · 106
参考文献 · · · · · · 107

7 学术论文写作中的问题汇总 · · · · · · 109

7.1 学术不端问题 · · · · · · 109
7.1.1 学术不端行为界定 · · · · · · 109
7.1.2 原因分析与写作建议 · · · · · · 110
7.2 论文结构与格式问题 · · · · · · 110
7.2.1 常见问题 · · · · · · 110
7.2.2 原因分析与写作建议 · · · · · · 111
7.3 论文内容方面的问题 · · · · · · 112
7.3.1 常见问题 · · · · · · 112
7.3.2 原因分析 · · · · · · 113
7.3.3 写作建议 · · · · · · 114
7.4 论文语言与文风问题 · · · · · · 115
7.4.1 常见问题 · · · · · · 115
7.4.2 原因分析与写作建议 · · · · · · 115
练习题 · · · · · · 117
参考文献 · · · · · · 119

附录 中文科技论文参考文献的运用及其著录规则 · · · · · · 120

参考答案 · · · · · · 133

1 总　论

学术论文的写作，一般提供给学术性期刊以期发表，或提交给学术会议，或是完成课程论文。其主要内容为报道学术研究成果，反映该学科领域最新的、最前沿的科学水平和发展动向。论文要有新意，即要求运用新观点、新方法以及新的数据或结论，兼具科学性。

1.1　学术论文英语写作的特征
（Features of Academic English Writing）

学术论文英语写作通常要求准确、鲜明、生动，应具备以下主要特征。

1.1.1　创新性

学术论文应是创新性研究的成果，其研究内容在写作之前没有人涉及过或涉及不深；要求运用新的理论或方法对他人已经研究过的方面进行重新研究、评估，或提供新的数据进行补充说明，或从全新的角度阐述，从而得出新的与众不同的结论。创新性是学术论文的生命。有创新才可以有独到的见解，才可以做到学术互动、理论互动、理论与实践互动，才能进一步推动学术攀登高峰，同时促进理论的完善和体系化，在实际生产中进一步节约成本、节省劳力、降低能源损耗，从而提高生产力。正是创新性使学术写作有别于一般性写作。

1.1.2　逻辑性

学术论文要体现学术性必须具有逻辑性。逻辑性包含两个方面的内容：观点清晰和陈述的连贯性。在整个写作过程中所讨论的主题应当协调一致，结构清晰，并富有创意。

1.1.3　启发性

学术论文写作的目的不是为了炫耀复杂的写作技巧，或故意使之深奥难懂而为难读者，因此要运用图表或数据，清楚生动地阐述相关结果或结论，而每一个观点都应当有具体的理论依据或细节。对观点的详细解释或对深奥的学术术语的解释有助于与读者有效的交流，从而达到启发读者的目的。

1.1.4　连贯性

连贯性指实用文中句子之间、段落之间衔接和过渡自然，从而增加阅读的流畅性和有效性。

1.1.5 准确性

准确性是指语言运用的正确和精准,以及对客观事物的运动规律和性质把握的接近程度。学术论文所使用的概念、定义、判断、分析和结论要准确,评价要适度,包括对研究成果的确切、恰当的估计和对他人研究成果实事求是的评价,切忌评价过度和评价不足而带来的片面性和歧视性。

1.1.6 规范性

学术论文写作的目的是学术交流和传播,启发他人思维。因此,论文的写作要遵循一定的模式和规范。文字表达要准确、简明、通顺,条理清楚,层次分明,论述严谨。对专业术语、数字、符号的使用,图表的设计和计量单位的使用,文献的引用,都必须遵循一定的规范。

1.2 学术论文英语写作的语言特征
(Language Features of Academic English Writing)

学术论文英语写作不同于一般写作,除一般学术性要求外,在语言和行文上也有自身的特点。

1.2.1 语言规范

学术论文通常使用正规的书面语,尽量不使用缩略语,不使用口头语言,包括词汇和语法。学术写作的目的是进行学术交流,对研究课题进行某一方面的探讨,介绍目前该课题的现状和存在的问题、所使用的研究方法、研究的结果或得出的结论。使用规范的语言有助于表明作者的严谨态度和对学术的尊重。具体体现在如下几个方面:

(1) 避免使用方言俚语,如 cool 或 pretty awful 等。

(2) 使用单词的完整形式,而不使用缩略形式,如 does not 和 it is,而不使用 doesn't 和 it's。

(3) 避免使用短语性动词,尽量使用单个的多音节词,如不使用 look into,而使用 investigate 或 examine。

(4) 避免使用不正规的标点符号,如缩略符号或破折号,而使用分号和冒号。

(5) 避免使用设问,尽可能使用其他修辞手法,包括消极修辞和积极修辞。

1.2.2 句子结构严谨

为了表达清晰的逻辑或推理,学术论文英语写作通常使用相对高级的句子结构,尤其是从句的大量使用成为写作水平的一个重要标志。英文从句通过连接词或关系代词、关系副词的使用能够把复杂的逻辑关系清晰呈现,同时突出主要内容,使行文流畅、有序,较好地表述各要素之间的逻辑关系。同时,学术论文英语写作还大量使用非谓语结构,如动名词、不定式和分词短语,使行文简洁,而逻辑关系蕴含其中。具体体现为:

(1) 较少使用并列结构,尤其是松散的并列,而更多地使用复合句。如:

The heat balance equation for each grid unit is found. The equation includes solar radiation, ground reflection, sky radiation, and etc.

点评：这个例子中的两个句子结构松散，连贯也不够，作为学术论文，必须提高表述水平，主要通过使用从句，才能更好地体现不同部分之间的逻辑关系。应改为：

The heat balance equation for each grid unit is found, which involves solar radiation ground reflection, sky radiation, etc.

又如：

Finally, the thermal characteristic of a certain reinforce computer case under the forced air convection are calculated. Some valuable results for the thermal design of electronic equipment are obtained.

点评：这个例子中出现了两个并列的被动句，但同样结构松散，如需提升语言表达层次，使其更为紧凑，可改为：

Finally, the thermal characteristics of a certain reinforced computer case under the forced air convection are concluded, with some valuable results for the thermal design of electronic equipment obtained.

（2）避免使用较短或较简单的句子，通常使用较长较复杂的句子。往往一个段落就只有一句话或两句话。更多情况下长、短句搭配，句式变换较多。如：

Movement from individual decisions to a group decision context often creates many economic ramifications, including strategic behavior and related game-theoretic effects. However, even apart from such influences, there are quite fundamental changes in individual choice behavior in terms of the information affection decisions, how people use the information available in group contexts to make decisions when preferences are noisy and how the group decision rule frames decisions in a manner that alters subsequent behavior. These alterations can have lasting effects on an individual subsequent decision.

点评：这个段落共有三句话，第一句使用了分词短语使句式相对简单，但这种写法恰恰是比较高明的；第二句非常复杂，从句套从句，是论文常用的句型；最后一句比较简单。这种长短、难易搭配使得句式变化较多，很能体现英语写作水平，从而体现论文水平。

下例很长很复杂，属典型的学术型英语句式：

While organizations are busy deploying eHealth information systems to better manage the quality and the delivery of health care services, from scheduling, billing, and health care records to the control of life-critical devices and clinical decision support, they face challenges that have to do with the overall complexity of healthcare, access to skills, and lack of interoperability among healthcare information systems.

点评：本句共61个单词。句子的主干是：they face challenges。while 引导让步状语从句，修饰整个主句。that 引导限制性定语从句，修饰先行词 challenges。整句话表达的意思是：尽管这些组织忙于部署电子健康信息系统，为的是更好地管理卫生保健服务的质量和投放市场，包括从编制目录、编制账单和卫生保健档案到生命危险期设置的控制和临床诊断支持，他们仍然面临着卫生保健信息体系中卫生保健、技能权限、协同性缺乏的整体复杂性的挑战。

1.2.3 大量使用专业词汇

学术论文英语写作的目的就是进行学术交流，是对某一专业性话题的深入探讨，其读者对象均为专业人士或专业学习者，而不是进行科学普及宣传，因此大量使用专业词汇就理所当然了，行话必然经常出现在行文中。专业词汇和行话对于普通读者来说可能有困难，但对于专业读者来说通常不存在困难。如：

To ascertain how much closed pores and coarse and fine pore channels influence the sintering of a sample pressed from the powder, we plotted Arrhenius dependences for the total volume and volumes of pore space and its components. For each curve, the slope to the x-axis, which is equal to the ratio of the difference in logarithmic volumes to the difference in reciprocal temperature, is proportional to the activation energy of decrease in the total volume or the volume of pore space or one of its components and may characterize the intensity with which the volume reduces. The dependence for the total volume has two linear regions corresponding to temperature ranges of active Line 1 and less intensive Line 2 sintering. Line 1 has greater slope to the x-axis than Line 2. Therefore, the total volume of the sample reduces much more intensively in the temperature range of active sintering than in the range of less active sintering.

点评：在本例中出现了大量的专业词汇，如 closed pores channels（闭孔流路），coarse pore channels（大孔流路），coarse and fine pore channels（大孔和精孔流路），Arrhenius dependences（阿伦尼乌斯函数变量），total volume（总量），logarithmic volumes（对数量），sintering（烧结）等。

1.2.4 学术英语的时态

学术英语大多使用一般现在时，客观地表示事实或通常情况下的状态或行为。实验过程偶尔采用过去时，但论述和结论部分都是一般现在时。在提及课题前景、今后要做的工作或可能产生的影响时，一般采用将来时。

（1）现在时（包括一般现在时和现在完成时）。介绍论文目的、研究方法、研究结论时采用一般现在时。

例1：In this paper, we study the dimensional effects, the electronic state, and the configuration of charge density and spin density for quasi-one-dimensional organic ferromagnets. （介绍研究内容）

例2：Compared to conventional parallel jaw grippers, multifingered robot hands have three potential advantages: they have a higher grip stability due to the multi-contact points with the grasped object; they can grasp objects of arbitrary shapes; it is possible to impart various movement onto the grasped object. In order for the multifingered grasp to possess the properties, it is necessary to study the grasp planning. The objective of this research is to develop algorithms of grasp planning and coordinative manipulation strategies for dexterous robotic hands. （介绍研究目的）

例 3：Because many important engineering surfaces are often surfaces with curvature, measurement and analysis of these curved surface topographies have become a frontier project in the field of surface metrology. On the basis of summarizing the current research situation and development of measurement surface topography, the author's research results of measuring theory and method for curved surface topography are discussed systematically. （提供依据）

例 4：The progress of science and technology is promoting the research in surface topography measurement. But so far the research contents at home and abroad have been focussed mainly on plane surface topography. （提及现状）

（2）过去时（包括一般过去时和过去完成时）。过去时多用于实验性论文，主要用于介绍实验过程和目的。在说明资料来源、收集方法、研究的起止时间、引用原始文献作者观点时，常用一般过去时。如：

例 1：This aim of our study was to perform such a prospective trial to determine whether there is a significant difference in the results of both the materials. （介绍研究目的）

例 2：In the second series of experiments, we created arrays of conical（圆锥形的）microstructures on the surfaces of silicon by irradiating a silicon wafer（晶片）with trains of ultrashort pulses in the presence of an ambient gas. The microstructures are typically tens of micrometers high and separated by several micrometers. We showed that the morphology（形态学）of the structures is highly dependent on the species of ambient gas, the gas pressure, the laser fluence, and the number of laser pulses used. The sharpest structures are made in haloger-containing（含卤）gases, such as SFs or C_{12}. The use of Ne or air results in blunt, rounded structures. （介绍研究过程）

例 3：This method resulted in a considerable reduction in pressure, as compared with those described in previous studies, without compromising the efficiency. （介绍研究结果）

例 4：From 2000 to 2010, the materials are applied in various experiments. （研究的起止时间）

例 5：In the first of the two experiments described in this thesis, we constructed and characterized an apparatus for performing ultrafast surface spectroscopy（光谱学）on Pt. （资料来源）

（3）将来时。将来时主要用于展望研究前景、介绍今后要做的工作、认定研究的价值和评估可能产生的影响等。有时为了委婉而使用过去将来时。如：

例 1：As interest in this area is increasingly intensified, more light will be thrown on this subject. （展望前景）

例 2：This research will produce significant implications for the applications of the theory in this area. （评价研究价值）

例 3：It would be helpful if the various ISO and IEC technical committees could be persuaded to define concepts in unison. （委婉表达）

1.2.5 学术英语的语态

合理的语态有助于提升论文的权威性和说服力。学术英语语态有两种，主动语态和被动语态。一般来说，这两种语态不可能独立出现，一般情况下同时出现，只不过在不同学科的论文中或在同一论文的不同部分所占比例不同。有时为了衔接和连贯，对主语进行变动，从而改变语态，甚至采用名词或名词性短语淡化语态，从而更好地实现连贯。

（1）主动语态。主动语态直接描述或说明主语的动作行为、存在状态等，有一个直接的行为执行者。在论文中需要突出研究者的努力和所取得的成绩时，使用主动语态。

例1：In this thesis, we discuss the application of these silicon microstructures in optoelectronics（光电子学）and vacuum microelectronics. We note that surfaces covered with these microstructures have striking optical properties: structures made in SFs absorb approximately 90% of incident light with wavelengths between 250 nm and 2.5μm.

点评：这个例子是主动语态，表明研究者所做的大量努力。

例2：We selected topics by reviewing the current scientific areas of interest at scientific meetings of British, European, and American hepatology groups. We also reviewed published reports by database and journal scanning and discussed ideas with colleagues.

点评：这个例子阐述了研究者已经完成的工作。

（2）被动语态。被动语态广泛运用于科技论文写作，因为被动语态能做到客观性，避免主观性，仅仅叙述事实，即在某人或某物上实际发生的事实或实际存在的状态，更符合科技论文对于客观性和科学性的要求。因此，被动语态大量运用于描述实验过程、实验结果以及一些客观存在的事实。

例1：Methods of evaluation are discussed and the results of the three land reclamation methods are compared and contrasted. Conclusions as to the effectiveness of the three methods are given, together with suggested recommendations for further studies.

点评：这个例子用被动语态描述了研究方法和结果。

例2：For the investigation, three differently remedied plots and one comparison plot were selected and the main physical and chemical properties of the soils were tested. The plot with the topsoil removed was still of high contaminants and low pH value. The plot remedied by using industrial mineral amendments showed a reduction in contaminants and a buffering of pH, but it still had a high level of metals. The subsoil covered plot showed the lowest content of contaminants, which was less than the threshold value.

点评：这个例子提供实验信息，使用了被动语态。

1.2.6 学术英语的人称

学术论文中人称的使用直接关系到论文的客观性和说服力。论文写作常用的人称为第一人称和第三人称。

（1）第三人称。一般来说，第三人称是论文写作必备的人称，广泛运用于各种论文，是运用最广、出现频率最高的人称。如下列各例：

例1：A combination of an optical parametric amplifier and a difference frequency generation stage allow us to create optical pulses with wavelengths in the range 550nm～7.3μm and pulse widths of approximately 100fs.

点评：文中使用了第三人称说明光参量放大器和差频率产生级组合的优势。

例2：The sharp conical structures also show high field-emission current. The remarkable and potentially useful optical and field-emitting properties of the structure are the results of the conical morphology of the laser-induced microstructures and the impurities introduced into the silicon during irradiation（辐照）.

点评：文中的 The sharp conical structures 和 The remarkable and potentially useful optical and field-emitting properties of the structure 两个第三人称说明了研究的结果。

（2）第一人称。有些学术论文中也大量出现第一人称，尤其是第一人称复数，这时第一人称的身份是观察者和记录者，不影响论文的客观性和说服力，有时提及实验者或研究者所做的前期准备工作和后期分析工作反倒能有效提升论文的客观性和可信度。很多论文在提到研究目的时都使用第一人称。

例1：We previously examined the morphology of tetragonal zirconia solid solution with transmission electron microscopy.（提及研究者所做的工作）

例2：In this article, we contribute to the understanding of the role that external risk perceptions play in decision regarding the combination of company-owned and franchised units in the hospitality industry, and to acknowledge of the impact of specific CEO characteristics on perceptions of environmental uncertainty.（提及研究目的）

1.2.7 口语体和书面体

例1：Nowadays companies are finding that they have to change the way they doings and they're finding that human resources planning is really helpful when they have to do this. One reason why it's helpful is because it can help the companies work out what the issues are and then, when you've done that, it can help you make up your mind what you're going to do about it. Basically, human resource planning is what you do when you're going through.（口语体）

例2：As companies experience the need for change, they often apply human resource planning to define the relevant issues and develop responses to them. Broadly defined, human resource planning is the process of analyzing an organization's human resources needs under changing conditions and developing the activities necessary to satisfy those needs.（书面体）

点评：同样内容的口语体和书面体差异极大，上面的口语体表达松散，不够规范，小词较多，多次使用缩略结构，并运用第二人称，这些都与学术写作相去甚远。而同样内容的书面表达结构紧凑、规范、简练，采用很多大词和专业词汇，并较好地运用词的搭配，显得地道而有深度。主动语态，现在时，但口语体同时运用第三和第二人称，书面体运用第三人称。

例 3: The rejection of Victorian gentility was, in any case, inevitable. The booming of American industry, with its gigantic, roaring factories, its corporate impersonality, and its large scale aggressiveness, no longer left any room for the code of polite behavior and well-bred morality fashioned in a quieter and less competitive age. War or no war, as the generations passed, it became increasingly difficult for our young people to accept standards of behavior that bore no relationship to the bustling business medium in which they were expected to ballet for success. The war acted merely as a catalytic agent in this breakdown of the Victorian social structure, and by precipitating our young people into a pattern of mass murder it released their inhibited violent energies which, after the shooting was over, were turned in both Europe and America to the destruction of an obsolescent nineteenth-century society.

点评：本例同样使用了规范的书面文体，各部分层次清楚，逻辑关系一目了然，并运用了并列手法和复合句，短短四句就有五个从句，且第三和第四句中从句套从句，但逻辑关系清清楚楚，并运用了主动语态、过去时和第三人称。

1.3 论文的组成部分

国际标准化组织，美国和英国的标准化组织都对科技论文的写作作出了规定。一些重要的学术期刊对所刊登的论文也有具体的要求。但是，一般说来期刊类科技论文主要由标题、摘要、正文、致谢、参考文献和附录等部分组成。

例如，The Proceedings of the Institution of Mechanical Engineers 对论文的内容和顺序要求如下：

The preferred order of contents is as follows：

（1）Title of paper（论文题目）。

（2）Author（s）name（s）and business address（es）（作者名字和机构名称）。

（3）Synopsis of not more than 200 words: covering the aim of the work, methods used, results obtained and conclusions reached, keywords for information retrieval purposes should be indicated（概括不超过 200 字：涵盖工作目的、使用方法、取得的结果和得出的结论，并注明信息检索的关键词）。

（4）List of notation in alphabetical order, defining all the symbols used in the paper（按字母顺序排列的符号列表，定义论文中使用的所有符号）。

（5）Body of the paper: organized into logical sections sequentially, numbered with no more than two grades of subheadings（论文的正文：按顺序分为逻辑部分，编号不超过两个级别的子标题）。

（6）Acknowledgements（致谢）。

（7）References in the order to which they have been referred in the text（参考文献按文中提及的顺序排列）。

（8）Appendices（附录）。

（9）Tables; these should be numbered consecutively throughout the text（表格；表格应在

整个文本中连续编号)。

(10) List of captions for the illustrations, which should be numbered consecutively throughout the text; both line drawings and photographs must be included in the same numbering sequence (插图的标题列表,应在整个文本中连续编号;必须在相同的编号顺序中包括线图和照片)。

1.3.1 标题和摘要

1.3.1.1 标题

标题是一篇论文给读者的第一印象,标题应该引人注目以吸引读者,但最重要的是标题要反映论文的主要内容。写标题时要选用:(1)最精确的词语;(2)表明论文要点的词;(3)有助于标引主题的词。

CBE Style Manual 对标题的写作提出了下述建议:

标题应在篇幅限制范围内尽可能提供信息,这应由期刊在其为作者提供的信息页中规定。标题应该是直截了当的描述性的。标题应该从一个词或术语开始,代表文章最重要的方面,下面的术语尽可能按重要性的递减顺序排列。正式科学术语的标准术语一般应该比普通或非标准术语更可取。

1.3.1.2 摘要

摘要的作用是:(1)迅速向读者展示全文是否值得进一步研究;(2)从完整报告中提取(抽象)以进行单独发布;(3)在个人或文献检索专家对索引和计算机库进行文献检索时,搜索术语来帮助文献检索。为达到这些目的,摘要必须是关于科学调查的简短、简明但不言自明的报告。

摘要的内容:摘要必须包括以下内容:(1)进行调查的研究目的和基本依据。(2)采用的基本方法。(3)研究结果和可以得出的有意义的结论。

摘要的长度:大多数期刊规定,摘要不应超过 200~250 个字,或论文本身长度的 3%~5%,表格应为一段。

CBE Style Manual 对摘要的写作提出下述建议:

研究报告的内容和顺序应准确、客观地反映文本,包括方法、结果和结论的主要要素。研究报告的摘要应当是信息性的(Ansi 1979),对所有内容要素作具体摘要。对于评论和其他类似的长而广泛的文章,摘要必须是指示性的(Ansi 1979),只是简单地勾勒出文章的主题,而不是总结证据和结论。除非单独理解缩略语,否则不应使用缩略语。摘要不应包括书目参考或列表数据。一般来说,摘要应该是单个段落,不带小标题。它不应超过期刊为作者提供的信息所规定的长度限制。摘要实例如下。

例 1:这是一篇研究圆柱形薄壳固有振动频率论文的摘要。

Prediction of natural frequencies for thin circular cylindrical shells

Abstract: In this paper an analytical procedure is given to study the free-vibration characteristics of thin circular cylindrical shells. Ritz polynomial functions are assumed to model the axial modal dependence and the Rayleigh-Ritz variational approach is employed to formulate the general eigenvalue problem. Influence of some commonly used boundary conditions, viz. simply supported-simply supported, clamped and clamped-free, and also the effects of variations of shell parameters on the vibration frequencies are examined. Natural frequencies for a number of particular cases are evaluated and compared with some available experimental and other analytical results in the literature on this topic. These results are given in the form of tables and figures. A very good agreement between these results of the present analysis and the corresponding experimental and analytical results available in the literature is obtained to confirm the validity and accuracy as well as the efficiency of the method.

Keywords: cylindrical shells, variations, boundary conditions, Rayleigh-Ritz method, Ritz polynomial functions, natural frequencies

参考译文：

圆柱形薄壳固有振动频率的预测

摘要：本文给出一种研究圆柱形薄壳自由振动特性的分析方法，用里兹多项式函数，模拟轴向模式的依赖关系，用瑞利里兹变分法构造一般特征值问题。文中研究了一些常用的边界条件，即简支—简支，固定端—固定端，固定端—自由端，以及壳体参数变化对振动频率的影响。对一些特定情况下的固有振动频率作了计算，并与试验结果以及有关本问题的文献中的其他分析结果作了对比。这些结果都用图表形式给出。本文的结果和相应的试验结果以及其他分析结果吻合得很好，表明本文的方法既有效又准确可靠。

关键词：圆柱形壳体，振动，边界条件，瑞利—里兹方法，里兹多项式函数，固有频率

1.3.2 正文的组成和写作

反映科研进展的科技论文可称为研究报告型论文（Research Report）。

它的正文包括 Introduction, Methods and Materials, Results, Discussion 等部分。各部分的功能如下：

Introduction describes the state of the knowledge that gives rise to the question examined by, or the hypothesis posed for the research. State the question or hypothesis.

Methods and materials describe the research design, the methods and materials used in the research (subjects, their selection, equipment, laboratory or field procedures), and how the findings were analyzed.

Results findings in the described research. Tables and figures supporting the text.

Discussion brief summary of the decisive findings and tentative conclusions. Examination of other evidence supporting or contradicting the tentative conclusions. Final answer. Implications for further research.

科技论文的目的是介绍作者的研究成果或研究进展，其内容反映作者对科学技术的发展所作出的贡献。对论文的读者来说，一篇科技论文的参考价值有三方面，一是论文的结论，二是研究的方法，三是论文提供的数据资料。撰写科技论文最重要的是论证是否正确，论点能否成立。因此，正文部分的写作应该特别注意论证过程的严谨，作到论点明确，概念准确，论据充分，推理过程符合逻辑。

为了达到这些要求，在写作时应注意以下几点：

（1）按逻辑关系组织文章的结构和段落，每段都用一个主题句开头。常见的段落展开形式有：从一般到特殊，从问题到解答，从原因到结果和按时间顺序。

（2）段落不要过长，每一段只述及一个问题。

（3）每一句只述及一个论点。不要把过多的内容塞到一个句子里，使读者不得要领。

（4）尽量使用简单的单词和短语，当有两个或更多的同义词可选择时，选用普通的、使用频率高的词。例如，表示期待可用 expect，也可用 anticipate，但以采用 expect 为好。

（5）缩写词在文中第一次出现时应给出其全称，以便于读者理解。

（6）对文中的重要概念，要自始至终用同一个词表示，在科技论文中不必苛求文采。

（7）用连接词或副词等清楚地指示出句子间的逻辑关系。

1.3.3 致谢和参考文献

1.3.3.1 致谢（Acknowledgements）

致谢放在正文之后，向论文所述研究提供资助的机构表示感谢，向对论文提供帮助的个人表示感谢等。

CBE Style Manual 指出：

An acknowledgements section can carry notices of permission to cite unpublished work, identification of grants and other kinds of finacial support, and credits for contributions to the reported work that did not justify authorship.

可用的句型有：

The research effort was partially sponsored by…

The authors would like to acknowledge the financial support of…

The authors are thankful to……for…

1.3.3.2 参考文献（References）

在论文中凡是引用他人著作中的数据、资料、方法、观点或结论的都应该标明，并在参考文献中列出来源。这一方面是对别人劳动成果的尊重，另一方面是提供研究工作的依据，便于读者全面地了解问题的起源和前人的研究工作。ASME Journal of Heat Transfer 对文献引用的标示和参考文献的书写格式规定如下：

Text citation. Within the text, reference should be sited in numerical order according to their order of appearance. The numbered citation should be enclosed in brackets.

Example：It was shown by Prisa [1] that the width of the plume decreases under these conditions.

In the case of two citations, the numbers should be separated by a comma [1,2]. In the case of more than two references citations, the numbers should be separated by a dash [5~7].

List of references. References to original sources for cited materials should be listed together at the end for this purpose. References should be arranged in numerical order according to their order of appearance within the text.

（1）References to journal articles, papers in conference proceedings, or any collection of works by numerous authors should include：

last name of each author followed by their initials；

year of publication；

full title of the cited article;

full name of the publication in which it appears;

volume number (if any) in boldface (Do not include the abbreviation " vol. " within the reference);

inclusive page numbers of the cited article.

(2) References to textbooks, monographs, theses, and technical reports should include: last name of each author followed by their initials.

year of publication;

full title of the publication (in italic or underlined);

publisher;

city of publication;

inclusive page numbers of the work being cited;

In all cases, the titles of books, periodicals, and conference proceedings should be underlined or in italic. A sample list of references follows.

外文作者姓名的著录格式采用姓在前（全拼，首字母大写），名在后（缩写为首字母），中间用空格；著作类文献题名的实词首字母大写，期刊文献题名的首词首字母大写，期刊名称请用全称，勿用缩写。具体如下：

(1) 单一作者著作的书籍：姓，名字首字母.（年）.书名（斜体）.出版社所在城市：出版社.如：

Sheril, R. D. (1956). *The terrifying future: Contemplating colortelevision.* San Diego: Halstead.

(2) 两位作者以上合著的书籍：姓，名字首字母., & 姓，名字首字母.（年）.书名（斜体）.出版社所在城市：出版社.如：

Smith, J., & Peter, Q. (1992). *Hairball: An intensive peek behind the surface of an enigma.* Hamilton, ON: McMaster University Press.

(3) 文集中的文章，如：

Mcdonalds, A. (1993). *Practical methods for the apprehension and sustained containment of supernatural entities.* In G. L. Yeager (Ed.), Paranormal and occult studies: Case studies in application (pp. 42-64). London: Otherworld Books.

(4) 期刊中的文章（非连续页码）如：

Crackton, P. (1987). *The Loonie: God's long-awaited gift to colourful pocket change?* Canadian Change, 64 (7), 34-37.

(5) 期刊中的文章（连续页码）：姓，名字首字母.（年）.题目.期刊名（斜体）.第几期，页码.如：

Rottweiler, F. T., & Beauchemin, J. L. (1987). *Detroit and Narnia: Two foes on the brink of destruction.* Canadian/American Studies Journal, 54, 66-146.

(6) 月刊杂志中的文章，如：Henry, W. A., III. (1990, April 9). *Making the grade in today's schools.* Time, 135, 28-31.

1.4 论文实例

为了对工程类论文整体有一个感性认识，下面给出一篇论文的框架，论文的内容是关于一种用于快速提取物体几何信息并将其传递给虚拟制造系统的方法。

An image-based fast three-dimensional modeling method for virtual manufacturing

Q Peng[1] M Loftus[2]

[1]Department of Mechanical and Industrial Engineering, The University of Manitoba, Winnipeg, Manitoba, Canada

[2]School of Manufacturing and Mechanical Engineering, University of Birmingham, UK

Abstract: This paper focuses on a quick way to extract geometric information from an object and transfer the information into a virtual reality (VR) system. Considering the problems associated with the construction of VR environments, different three-dimensional modeling methods are analyzed in this paper, and the authors report an effective methodology to construct virtual manufacturing environments. A three-dimensional surface reconstruction system, based on image analysis, forms a new approach to data acquisition and promises to be a competitive technique for VR applications. This three-dimensional surface reconstruction system integrates the binocular stereo principle and the shape from shading technique. Experimental results are encouraging and the performance of the system has been evaluated by simulation procedures.

Keywords: virtual manufacturing, CAD, reverse engineering

Notation

b: displacement between two cameras used for obtaining the image pair

f: focal length of the camera

I_n: incident intensity of the point light source

I_{co}: image coordinate system

K_d: constant diffuse reflectivity

N: surface normal

p, q: parital-derivatives of height Z with respect to image coordinates

p_s, q_s: partial derivatives of the light direction

$r_{woo} = (x, y, z)^T$: coordinate vector of a point relative to the world coordinate system

$r_{Ico} = (x, y, z)^T$: coordinate vector of a point relative to the image coordinate system

R_{Ico}: roatation matrices between the image coordinate and the world coordinate

T_{Ico}: translation matrices between the image coordinate and the world coordinate

W_{co}: world coordinate system

$x_l - x_r$: pixel shift at a point between two corresponding images

$z = \zeta(x, y)$: surface function of the object
$\theta(x)$: Heaviside function
λ: Lagrangian multiplier used to enforce the constraint of surface smoothness
$\zeta(x, y)$: reflection coefficient
Ω, ω: thresholds

1 Introduction

The exciting new technology of virtual reality (VR) is being increasingly employed to improve and the reduce the costs in design and manufacturing. The great potential of VR is that it for the design and manufacturing process. However, this places a heavy demand on the geometric modelling task in the virtual design and manufacturing system, and building an environment around a particular system continues to prove one of the most difficult development problems. For instance, a replica of a factory floor with its machine tools and supporting equipment could be modelled on a computer aided design (CAD) package. Each item would have to be individually created as a three-dimensional model and merged together to form the whole manufacturing environment. This accounts for a significant portion of the overall process.

Therefore, it is a challenging task to create a completely new virtual environment and it needs support from efficient three-dimensional modeling methods. Most of the current virtual CAD/CAM (computer aided design/manufacture) systems use a "synthetic" virtual world that is based on advanced computer graphics and realtime simulation technology[1~4] as their virtual environment.

The first step to set up a virtual "synthetic" environment is to make a copy of the real world. The process of building a virtual environment is shown in Fig. 5.2. In each loop, the system gathers information about the environment from sensors, such as a mouse, spaceball or camera, and processes the input data to form the shape of the object and update the VR environment.

Current commercial VR software provides powerful R functionality but may fall short of creating the new VR environment. The shape information of a new object has to be built by other CAD software, such as Pro/Engineer or AutoCAD, in a standard CAD format and then translated into a virtual object in a VR system. An alternative method is to use virtual reality modeling language (VRML) to download existing three-dimensional object models from three-dimensional shape libraries by Internet. Using a CAD system to define the required three-dimensional shape can substantially reduce the shape building work, but building a virtual world still requires significant time, especially when CAD libraries are not available. The most general demand of a VR system is to make users feel that they are a part of the environment and can easily interact with it, ie. provide an immersive capability.

The approach to building a VR environment should ideally be:

(1) Fast. It should catch and build the object quickly and meet the real-time requirement.

(2) Automatic. The process should be able to extract all the shape information without input from an operator.

(3) Versatile. The method should not be limited to simple geometries but should be able to

handle complex items.

(4) Accurate and efficient. Different three-dimensional modeling methods, such as CAD/CAM system-based methods, three-dimensional scanner-based approaches, magnetic sensor-based methods and image-based approaches are analyzed in this paper, and the authors report an effective methodology to construct virtual manufacturing environments. The main aim of the methodology is to extract the geometric information from a three-dimensional object and transfer this information into the VR system. The new three-dimensional surface construction approach integrates the binocular stereo principle and the shape from image information technique. The experimental results are encouraging and support the development of VR technology and its applications into the manufacturing environments.

2 Three-dimensional shape capturing methods

Several methods which can provide a quick means to construct VR environments are being considered.

(1) CAD/CAM system-based methods. An integrated CAD/CAM system allows users to design the manufacturing process as a unified model representation that supports the transfer of the model between software systems.

Feature-based methods are used by most of the current CAD systems. The feature extraction and recognition method analyses the geometry and topology of the existing CAD model and attempt to classify geometric features of the model as different manufacturing entities[5]. This method is limited to the existing CAD model. Starting with a simple shape, features are added or modified to refine the model. This approach speeds up the model creation process because previous designs can be used as the basis for new models. However, this restricts the user using new features to define the model. It will also take a long time to build a new VR environment from the start if an existing CAD database is not available. This has forced researchers to look to other emerging technologies.

When forming an immersive three-dimensional environment[6,7].

(2) Three-dimensional scanner-based methods. The three-dimensional laser scanner is a popular and accurate method of recovering three-dimensional data from an object. However, its capturing process is slow, because it scans the surface line-by-line and point-to-point. The high energy light source also needs to be treated with care.

An alternative approach to three-dimensional shape capture is to use tactile methods that measure the surface of an object by the touch of mechanical arms. The relative coordinate positions are determined by the sensors in the joints of the arm, which can be supported by robotic devices.

This method is accurate and reliable but the slow speed is a disadvantage[8].

(3) The magnetic sensor-based technique. In the magnetic sensor system, a transmitter is driven by a pulsed d.c. signal, as shown in Fig. 5.3. The sensor measures the transmitted magnetic-field pulse referenced to the magnetic transmitter. An electronics card controls the transmitting and receiving elements and converts the received signals into three-dimensional position and orientation measurements. This method supports realtime measurement with high accuracy, but its

disadvantages include the touch measuring process with wire and sensor, limited captured points from the sensors and the limitations of the magnetic-field that is applied.

(4) Image-based approaches. Image-based methods have the special features of noncontact measurement, low energy and capturing all the surface data at the same time. The ultrasonic imaging system consists of a transmitter, an ultrasonic receiver array, and an object to be recovered. The three-dimensional surface of the object is defined by $f(x, y, z) = 5(x, y) e^{(5(x,y)-z)}$.

Therefore, the three-dimensional shape of the object can be reconstructed from the observed pressure of the scattered wave at specific locations and time[9].

Even in dark environments, the images of an object can be obtained by the ultrasonic system. However, it has seldom been used for surface reconstruction because the following factors lower the resolution: long wavelength, limited number of receivers, attenuation due to propagation through air, and the non-linear response of the ultrasonic receiver. An alterative way is to use charge coupled devices (CCDs) or digital cameras as the sensors to capture the image, and apply the approaches, of image processing to recover the object information.

The methods of three-dimensional shape reconstruction, based on vision information from a CCD or a digital camera, have received considerable attention and various methods have been used to capture the three-dimensional depth information of objects from their two-dimensional digital images[10~13]. These methods can be divided into two groups: photometric stereo-based and binocular stereo-based. The photometric stereo-based techniques use images of the same scene with different light source directions. If the surface reflectance properties of the object are known, the local surface orientation and the height of illuminated points can be computed by the techniques of shape from shading. One of the important advantages of this method is that it does not suffer from the correspondence problem, which is referred to later in this paper. It has been used effectively by the authors in reverse engineering applications[14]. Unfortunately, as the images are captured at different times, this approach cannot meet the needs of real time in a VR system. The binocular stereo-based techniques extract the depth information from intensity images acquired by two cameras at the same time; the two cameras being displaced from each other by a known distance. The key to this method is to determine the point in one image which corresponds to a given point in the other image, namely the correspondence problem. A significant amount of work has been carried out to solve this challenging problem, but the results have not been satisfactory[11], because time delays hinder real-time applications. The technique is also limited by the sparse (feature points) depth information which it can provide.

This paper outlines an approach to overcome these problems by integrating the information from the shape from shading process with the binocular stereo-based technique. The three-dimensional shape information of an object is recovered from the key feature points extracted by the binocular stereo-based method and the shape from shading method reconstructs the surface of the object.

3 The shape from shading method

In order to describe the difficulties in recovering a three-dimensional shape from an image, it is necessary to understand the nature of reflection from an object and the associated image information process. Most implementations of the shape from shading module assume a Lambertian surface, which only considers diffuse reflection from materials with a distance point source.

Given a Lambertian image, the image irradiance equation relates the intensity of an image point (x, y) with the surface normal N, as follows[10]:

$$I(x, y) = R(p, q) = k_d I_n L \cdot N$$
$$= k_d I_n \frac{1 + pp_s + qq_s}{\sqrt{1 + p^2 + q^2}\sqrt{1 + p_s^2 + q_s^2}}$$

Shape from shading by a single image does not produce accurate results. The surface normal at each point is represented by two numbers, p and g, the two variables representing the surface normal direction at each point cannot be computed by using one equation. Boundary conditions and local constraints based on reasonable assumptions support the most common reported solutions.

A direct formulation of the shape from shading problem, with a smoothness condition on the surface, is stated in the following minimization problem:

$$\iint_\Omega \left\{ \left[I(x, y) - k_d I_n \frac{1 + pp_s + qq_s}{\sqrt{1 + p^2 + q^2}\sqrt{1 + p_s^2 + q_s^2}} \right]^2 + \lambda (p_y - q_x)^2 \right\} dxdy \rightarrow \min$$

where $p = q$. Therefore, p and g can be computed by the iterative scheme from equation, as follows[10]:

$$p_{ij}^{n+1} = \bar{p}_{ij}^n + \lambda [I_{ij} - R(p_{ij}^n, q_{ij}^n)] R_p(p_{ij}^n, q_{ij}^n)$$
$$q_{ij}^{n+1} = \bar{q}_{ij}^n + \lambda [I_{ij} - R(p_{ij}^n, q_{ij}^n)] R_q(p_{ij}^n, q_{ij}^n)$$

where

$$\bar{p} = \frac{1}{2}(p_{ij+1} + p_{1j-1}), \quad \bar{q} = \frac{1}{2}(q_{i+1j} + q_{1-1j}), \quad R_p = \frac{\partial R}{\partial p}, \quad R_q = \frac{\partial R}{\partial q}$$

Unfortunately, even for ideal Lambertian diffuse reflectance, this model has practical difficulties in achieving an accurate solution for three-dimensional shape recovery. However, the shape from shading process can converge to a unique solution when the boundary condition is available. The binocular stereo-based method can be used to find this information.

4 The shape from binocular stereo method

...

5 The integration of the binocular stereo and shape from shading methods

...

6 Three-dimensional environment capturing and accuracy analysis

...

7 Conclusions

Many of the ideas and technologies of VR are still being researched, but progress suggests that VR technologies will play a key role in future design and manufacturing application. The initialresults from the suggested method are encouraging and support VR environments to be constructed with little expense and effort. The approach described in this paper is still under development in the following directions:

(1) To develop an automatic process system to link a group of images in order to form a whole recovery shape of an object, because the pictures taken from one angle can only be used to recover the visible surface of the object.

(2) A fast calibration system is needed to cooperate with the camera system in order to build three-dimensional model rapidly and efficiently.

(3) Colour image information is to be used to improve the recovery accuracy.

Acknowledgement

The authors acknowledge the support given by the CVCP for the ORS award and the School of Manufacturing and Mechanical Engineering at the University of Birmingham for additional assistance.

References

[1] Kim S, Sonthi R, Gadh R. Virtual engineering environment applications: an over-view [J]. In Design for Manufacturability, 1994, 67: 1~11 (American Society of Mechanical Engineers, New York).

[2] Lefort L, Kesavadas T. Interactive virtual factory of a shopfloor using single cluster analysis [C] // In Proceedings of the 1998 IEEE International Conference on Robotics and Automation, Leuven, Belgium, 1998: 269~271.

[3] Chuter C, Jernigan S R, Barber K S. A virtual environment simulator for reactive manufacturing schedules [C] // In Proceedings of Conference on VR in Manufacturing Research and Education, Chicago, 7~8 October, 1996 (American Society of Mechanical Engineers, New York).

[4] Mourant R, Wilson B H, Li M, et al. A virtual environment for training overhead crane operators [C] // In Proceedings of Conference of VR in Manufacturing Research and Education, Chicago, 7~8 October, 1996 (American Society of Mechanical Engineers, New York).

[5] Kraftcheck J, Dani T, Gadh R. State of the art in virtual design and manufacturing [J]. VR News, 1997: 6 (4), 16~22.

练习题

1. You can see an incomplete abstract of an article with the title. "Margaret C. Anderson's Litle Review", which has only the beginning and the end. Please put following sentences marked with A, B, C and D in the right order to make a complete abstract. This research looks at the work of Margaret C. Anderson the editor of the Litle Review. ①②③④This case example shows how little magazine publishing is arguably a literary art.

A. This research draws upon mostly primary sources including memoirs, published letters, and a complete collection of the Litle Review.

B. This focus undermines her role as the dominant creative force behind one of the most influential litle magazines published in the 20th century.

C. Most prior research on Anderson focuses on her connection to the famous writers and personalities that she published and associated with.

D. The review published first works by Sherwood Anderson, James Joyce, Wyndham Lewis, and Ezra Pound.

2. The following abstract passage is incomplete with several key words missing. Please fill in each blank with a proper word from the box below.

| application | applying | assessed | benefits | conclusion | developed |
| research | showed | therefore | suggestions | summary | judge |

Lasers have been used in the entertainment industry since 1964, when they were used in the film Goldfinger. Laser display shows commenced in about 1973. It would be reasonable to expect laser safety to have been adequately addressed over the last twenty-five years. This research ① that the industry was not able to assess the risks from its work. A national survey of the competence of enforcing officers showed that they rarely had the necessary expertise to ② the safety of shows. ③, there was often a wide gulf between the laser companies and those responsible for enforcing entertainment and health and safety legislation.

A hazard assessment methodology has been ④ which considers any laser show as a series of modules which may have different hazards associated with them at different stages of the life cycle, and different people would potentially be exposed to these hazards.

A number of laser radiation exposure situations have been ⑤ including audience scanning. A theoretical understanding of the laser scanning issues and the ⑥ of measurement techniques to enable assessments to be carried out against internationally recognized maximum permissible exposure levels were developed. The ⑦ was that the practice of audience scanning was not acceptable in its current form. A number of laser companies worldwide have accepted this view as a direct result of this ⑧.

A means of presenting the risk assessment for a laser display has been developed which provides ⑨ for the laser company, the venue manager, event promoter and the enforcing officer. It is recognized that a complete assessment may not be possible in the time available and a focused approach to the assessment is presented. In ⑩, if audience scanning is intended, the assessment is presented. If audience scanning is intended, the assessment is complex, but if this practice is not intended then the assessment can be straight forward.

⑪ are made for ⑫ the risk assessment methodology to other laser applications.

3. The following introduction is a rough draft full of mistakes. Please mark them and insert proper corrections between the lines.

Technological advances in the 1990s started new ways of conducting business in the US and throughout the world. Very important was the World Wide Web, the Internet allowed consumers and businesses to communicate in completely new ways that were previously unknown. From the mid-1990s to early 2000, the focus of businesses was about the opportunity given by the new Internet. Following the "doc-com" crash in early 2000, however, businesses began to know the problems with doing business on the Internet. From this point on, much of investors' and the media's focus shifted to companies with both Internet and "bricks and mortar" presences.

Many researchers have investigated on-line buying in the past several years (e.g. Jarvenpaa & Todd,

1997; Lohse, et al., 2001). However, little research deals with the relationships between on-line purchasing ("clickes") and purchases made at physical stores ("bricks"). The purpose of this paper is to report the results of two studies to evaluate consumers' reactions to shopping at clicks and bricks. The paper first review background relevant to the two shopping channels and the challenges faced by companies operating the two shopping channels and the challenges faced by companies operating in the on-line environment. Then we develop a research model and research questions to guide our investigations. Then we report the results of a large survey aimed at uncovering several of the relationships specified in the model, and study of a single retailer operating in both the clicks and bricks environments. These studies provide insights into consumers' experiences with shopping on-line and at physical locations, as well as to guide retailers take maximum advantage of the two shopping channels.

参考文献

[1] Hashimoto I Y, Kroll B M, Schafer J C. Strategies for Academic Writing [M]. Ann Arbor: The University of Michigan Press, 1985.

[2] Laster A A, Pickett N A. Occupational English [M]. New York: Harper&Row, 1991.

[3] Lindsay W M. Guide to Scientific Writing [M]. London: Addison-Wesley Longman, 1995.

[4] Silyn Roberts H. Writing for Science and Engineering: Papers, Presentations and Reports [M]. Oxford: Butterworth-Heinemann, 2000.

[5] 胡庚申. 英语论文写作与发表 [M]. 北京: 高等教育出版社, 2000.

[6] 罗立胜. 同等学力人员申请硕士学位英语统考指导丛书——写译分册 [M]. 北京: 中国人民大学出版社, 1999.

[7] 秦伯益. 如何写论文做报告 [J]. 中国研究生, 2003 (5).

[8] 秦荻辉. 实用科技英语写作技巧 [M]. 上海: 上海外语教育出版社, 2001.

[9] 唐国全, 等. 大学英语六级考试——阅读手记 [M]. 北京: 航空工业出版社, 2002.

[10] 田力平. 论文写作与网络资源 [M]. 北京: 北京邮电大学出版社, 2002.

[11] 王建武, 李民权, 曹小珊. 科技英语写作——写作技巧·范文 [M]. 西安: 西北工业大学出版社, 2000.

[12] 叶云屏, 孙逢春. 实用英语写作新途径 [M]. 北京: 国防工业出版社, 2001.

2 标题和关键词

2.1 概 述

2.1.1 标题的定义

科技论文的标题（Title）是能表达文章的特定内容，反映文章的研究范围和研究深度，是文章基本思想的浓缩与概括。为了使文章赢得相关领域的读者关注，论文的标题必须用最精练的语言恰如其分地体现全文的主题和核心。因此，论文的标题要求简明扼要，突出主题思想，表达论文的中心意思。从语言的角度来讲，标题往往是名词性词组，多为中心词+修饰语，或采用分词的形式，句式一般是中心词在前，修饰词在后，一般不采用句子的形式。

2.1.2 标题的作用

标题的作用在于吸引读者、帮助文献追踪或检索。标题相当于文章的"标签"（Label），一般的读者通常是根据标题来考虑是否需要阅读摘要或全文，而这个决定往往是在一目十行的过程中做出的。因此，标题如果表述不准确，就会失去其应有的作用，使真正需要它的读者错过阅读的机会。此外，文献检索系统多以标题中的关键词作为索引，因而这些词必须要准确地反映文章的核心内容，否则就有可能产生遗漏。

2.1.3 标题的格式

标题应用有限的字数概括全文的主旨，用词必须仔细斟酌和推敲，选择最简练、最准确、最贴切的词来表达全文的主要内容。为方便检索，标题还应尽可能的包含论文的关键词。标题要简明精练，不宜超过 20 个字，应删去无实质性内容、可有可无的词语。如"关于……""试析……"等词一般可以省去；没有特定定语成分的"研究""分析"等，也应被视为赘词，予以删除。此外，标题不易使用外来语、缩写词、简称、符号等，应该尽可能的使用本学科普遍接受、词义单一、便于使用的规范术语。

2.1.4 关键词的定义

关键词（Keyword）用来表达文章主题内容及信息目的，由单词或术语构成，可以直接从文章中选取，是表达全文主题内容最关键的词语，是文章思维方法的提炼和概括。也可以是经过反复推敲锤炼的词，不是词组或句子。还可以是术语，即某个学科中的专业词语。

2.1.5 关键词的作用

关键词的主要作用是通过词的逻辑组合,准确揭示文章的主要内容。有的关键词可以直接从标题中摘取,但也不是所有的关键词都能够从标题中提取。为了提炼出最核心、最主要的反映文章主题内容的词,也可以从摘要中摘取,有时还需要透析全文,直至提取出能反映主体内容的词,使读者对文章的主体结构及其价值很快做出判断。

2.1.6 关键词的格式

关键词应具有代表性、专指性、规范性和可检索性,其功用主要是便于检索,了解论文主题和输入计算机检索系统。科技论文的关键词应为 4~8 个。第一个关键词列出论文主要工作或内容所属二级学科名称,如:道路工程、桥梁工程、隧道工程、交通工程、汽车工程、机械工程、交通运输经济等。关键词的选择应与论文主题一致,能够概括其内容的词和词组,使读者通过关键词即可大致判断出论文的研究对象、材料、方法、过程和条件等。

2.2 基本要求

2.2.1 标题的简洁性要求

标题要准确地反映文章的内容,既不能过于空泛和一般化,也不宜过于繁琐,得不出鲜明的印象。标题中的术语应是文章中重要内容的亮点,并且易被理解和检索。标题应尽量避免使用非定量的、含义不明的词,如 "great" "fast" 等。

(1) 不宜过长,简洁精练。英语科技论文中,标题不宜过长,大多为 8~12 个单词左右。论文的标题强调语言精练、文字简短、用词精选,应能反映该文章所涉及的学科和所要研究的核心内容。若简短标题不足以显示文章内容或反映出属于系列研究的性质,还可使用正、副标题,以加副标题来补充说明特定的实验材料,方法及内容等信息。例如,Discussion on Several Academic Problems in the Book *The Modern History of Chinese Architecture* 的标题用词近 15 个,而 discussion on, the book, academic 等基本起不到核心作用。从分析来看,写作学术文章的目的就是探讨学术问题,进行学术研究,故 academic 无用。书名已按斜体字母以及标题和实词的首字母大写标出,故 the book 无用。新的学术问题要通过讨论、证明从而扩大已知领域,故 discussion on 也无用。删除冗赘无用部分后符合标准的标题应为: *Several Questions in Modern History of Chinese Architecture*[1]。

(2) 表达清晰,重点突出。标题要清晰地反映文章的具体内容和特色,明确表明研究的独到之处,力求简洁有效、重点突出。为表达直接、清楚,以便引起读者的注意,应尽可能地将表达核心内容的主题词放在标题开头。如:The effectiveness of reinforcement and reconstruction in steel-structured buildings 要比把主题词 reinforcement and reconstruction 作为标题的开头合适。因为用 effectiveness 作为标题中第一个主题词,就直接指明了研究问题:

---

[1] 张超英,等. 科技论文英文题名制作的技术规范 [J]. 编辑学报,2006 (12): 14~16.

Is reinforcement and reconstruction in these buildings effective? 表意不清的标题往往会带来困惑，如：The effects of oxidant on its running 中的 its 表意不清。

标题中应避免使用化学式、上下角标、数字符号、希腊字母、公式、不常用的专业术语和非英语词汇等。

2.2.2 标题的句法要求

（1）多用名词及名词性结构。标题中用得最多的是名词（包括动名词），有的标题中 80%以上的词为名词。由于标题是对全文重要内容的高度概括，因此用词要贴切、中肯，不能有任何随意性。如在标题 Temporal and spatial growth of wind wave 中，从句法层面上看，growth 是中心词，前面两个形容词 temporal and spatial 和后面的 of 介词短语都是它的定语。从语义或信息层面上看，growth 又是语义中心或信息中心，是标题要突出的焦点。这种两个中心合二为一的表达可以使标题结构紧凑、信息焦点突出。❶

名词性词组由名词及其修饰语构成。名词的修饰语可以是形容词、介词短语，有时也可以是另一个名词。名词修饰名词时，往往可以缩短标题的长度。以下各标题分别由两个名词词组构成。例如：

Soil Behavior（名词词组）and Critical Soil Mechanics（名词词组）

High Speed Flow Sensor（名词词组）and Fluid Power Systems Modeling（名词词组）

Traditional Versus Adult Studies Students（名词词组）：the College Experience（名词词组）

（2）正确使用介词以及介词词组。标题中除名词外，用得较多的是介词。介词在英语中是很活跃的一类词，在英文标题中"of""for""in""on"的使用频率较高，它们虽然都可引短语作定语，但在含义上还是有差别的。"of"侧重于所属关系，"for"表对象、目的，"in"表范围。

介词词组由介词+名词或名词词组构成。如果整个标题就是一个介词词组的话，一般这个介词是"on"，意思是"对……的研究"。例如：

On the Distribution of Sound in a Corridor

On the Crushing Mechanism of Thin Walled Structures

（3）名词/名词词组+介词词组结构。这是标题中较为常用的结构。例如：

Fundamentals（名词）of Flow Measurement（介词词组）

Scattered Sound and Reverberation（名词词组）on City Streets and in Tunnels（介词词组）

Acoustics（名词）of Long Spaces（介词词组）：Theory and Application（名词词组）

Investigation（名词）of Air Bags Deployment Forces（介词词组）on Out-of-Position Occupant（介词词组）

标题中的介词词组一般用来修饰名词或名词词组，从而限定研究范围。这种结构与中文的"的"字结构相似，区别是中文标题中修饰语在前，中心词在后。英文正好相反，名

---

❶ 李平，曹雁. 科技期刊论文英文标题抽象名词短语结构分析与应用 [J]. 中国科技期刊研究，2012，23（2）：322~324.

词在前，而作为修饰语的介词短语在后。

2.2.3 标题的指示性要求

标题通常使用指示性语言而不是报道性的，它说明文章的主题，而不陈述结论。指示性标题通常不是一个完整句子，不带谓语，所以很少使用完整句子作为标题。例如：

Migrating fields respond to radar electromagnetic fields（移动范围取决于雷达电磁场）中 respond 是动词，充当谓语，标题是一个完整的句子。

通常情况下，使用名词性词组表示指示性标题。例如：

Effect of radar electromagnetic fields on migrating fields（雷达电磁对移动范围的影响）

又如，很少用下面题目：

Short titles are often easier to read（简短题目往往读起来容易些）

而常用下面题目：

Relationship of brevity and readability in titles（题目简短性与可读性关系）

科技论文的中文标题往往是由表示并列或动宾关系的词构成的复杂词组，而英文标题几乎全部是以名词作中心词，用介词短语、名词、形容词、分词、不定式或从句加以修饰的复合词组。❶

2.2.4 标题的朴实性要求

科技论文题目注重科学性，不强调艺术性，要避免使用隐喻、比喻等修辞手法，还要避免使用俗套语（Formulas）、土话、隐语（Jagon）、多余的形容词和华丽的辞藻、广告式的渲染夸张词汇、感叹词、象声词及其他带有浓厚感情色彩的词汇。题目不要带惊叹号（Exclamation）或问号（Question mark）。虽然有论文题目以问题形式呈现，如："Stability Analysis: Where do we stand?"，但不常采用。

2.2.5 标题中尽量使用关键词

标题中的用词多是文章的关键词，明确、精练，将文章的主要内容予以高度概括。标题中用得最多的是名词或名词词组，一般不用动词或动词词组。如果用动词，则用非谓语动词形式，如动名词、不定式或分词。例如：

标题 1：On the Fatigue Life Prediction of Spot Welded Components

关键词 1：fatigue, spot-weld, automobile, life prediction

标题 2：Computer Simulation and Experimental Study On Cold Shut During Mold Filling

关键词 2：mold filling, computer simulation, cold shut, casting

标题 3：Investigation of Air Bags Deployment Forces on Out of Position Occupant

关键词 3：air bag, out-of-position occupant

标题 4：New Fatigue Test and Statistical Method for Metallic Materials Used in Vehicle Transmissions

关键词 4：fatigue test, statistical method, test specimens

---

❶ 陆艾五. 科技论文的中英文标题写作［J］. 安徽农学院学报，1990，10（2）：131~132.

文章中的关键词语被在标题中使用，不但有助于概括文章的主旨并减少标题中无用词语的数量，还可增加被检索、被引用的次数。

2.2.6 关键词的选取要求

关键词是从题目和文章中选出来，用于表达文章主题内容的单词或术语。关键词主要通过词的逻辑组合，精确揭示文章的主要内容。一篇文章可以选取 3~8 个关键词，可以采取直接从题目中摘取，也可以提炼反映文章主题内容的词，或从摘要中摘取，有时还需要分析全文，直至提取能反映主体内容的词，以便对文章的主体结构及其价值做出判断。[1]

关键词一般为名词，多是单词和词组而非缩略语。没有检索价值的词不能用作关键词，如"技术""应用""观察""调查"等。化学分子式不作为关键词使用，缩写词不能作为关键词。科技文章中常提到的技术，不能作为关键词。每篇文章的关键词一般为 3~5 个，最多不超过 10 个。英语关键词要与中文关键词相互对应且数量一致。

2.3 写作范式

2.3.1 标题的构成

标题通常由名词或名词性短语构成，如果必须使用动词，则多为分词或动名词形式。由于标题主要起指示性作用，而陈述句容易使标题具有判断性的语义，所以陈述句一般不易作为论文题目。例如："Fruit Flies Diversify Their Offspring in Response to Parasite Infection"中动词 diversify 的出现，使其看上去像某种断言。

有时可以用疑问句作为标题，尤其是在评论性文章的标题中，使用含有探讨性的疑问句型标题显得比较生动，易引起读者的兴趣，但这种用法不常使用。例如："When is a bird not a bird?""Should the K-Ar isotopic ages of olivine basalt be reconsidered?"等，生动且切题。

2.3.2 标题的句法规则

（1）译成名词短语。

1）基于 DSP 与 PGA 的实时数字信号处理系统设计：Design of real-time digital signal processing system based on DSP and FPGA。

2）由原噪声干扰对合成孔径雷达干扰效果的仿真研究：Simulation of the active noise Jamming's effect on SAR。

以上两个标题分别以名词性短语提炼出句子主干，使人一目了然。

3）齿刀是怎样演变成齿轮滚刀的：Transformation from rack-type gear cutter into gear hobber。

该题目中文表达为疑问句式，但如果将相应的英文标题也译为疑问句就不甚恰当。

---

[1] 杜兴梅. 学术论文摘要与关键词的写作及其格式规范 [J]. 韩山师范学院学报，2008，29（2）：82~87.

(2) 正确使用介词，常用介词短语形式。科技论文英文标题中常用介词短语结构，如介词"on""about"等。如：

1) 基于归结原理的 XML 问题求解：On XML problem solving based on resolution principle。
2) 浅谈多媒体技术：About multimedia techniques。
3) 论脉冲位置调制组和雷达信号：On the PPM hybrid radar signal。

介词"on""about"常用于标题表示"有关""论"等含义。

介词"for"有表示目的、用途之意"为……"，"in"表示方位"在……"。如：

1) 复杂电子系统测点与诊断的优化方法：Efficient algorithm for test-point and diagnosis strategies of a complex electronic system。
2) 一种低信噪比下点目标检测新方法：A new point target detection algorithm in low SNR。
3) 一种具有前馈补偿的滑模鲁棒控制器设计方法：A design method for sliding mode robust controller with feed forward compensator。

（资料来源：唐耀，2005）

(3) 避免用词累赘。用词多余也称作用词累赘，通常是将几个意义相近或不需表达出的词罗列在一起表达，从而造成意思重复。如：

1) The investigation *work*…
2) *In the field of* earth science…
3) *Research on* XML problem solving based on resolution…

以上斜体处可以不要，对句义无影响。

(4) 标题中单词的大小写。标题中字母主要大写有三种形式：全部大写、首字母大写、每个实词首字母大写。对于专有名词首字母、首字母缩略词、句点后任何单词的首字母等在任何情况下均应大写。

2.3.3 标题大小写错误

(1) 单词大小写常常混用。英文题目的大小写一般有以下 3 种格式：字母全部大写；每个单词的首字母大写，但 3 个或 4 个字母以下的冠词、连词、介词全部小写；题目第 1 个单词的首字母大写，其余字母均小写。例：Microelectronic assembly and packaging Technology 句子当中"Technology"不是特定单词，但是其首字母却用了大写。正确的写作形式应该为 Microelectronic assembly and packaging technology。

(2) 化学元素符号不大写。化学元素的符号可以用其拉丁文（或英文）的第一个大写字母来表示，例如：氮的拉丁文是 Nitrogenium（英文是 Nitrogen），其元素符号为"N"；氧的拉丁文是 Oxygenium（英文是 Oxygen），其元素符号为"O"；氢的拉丁文是 Hydrogenium（英文是 Hydrogen），元素符号为"H"。若化学元素的拉丁文（或英文）的首字母相同，需要补充添加一个小写字母以区分，例如：碳（拉丁名为 Carbonium，英文为 Carbon）用"C"表示；铜（拉丁名为 Cuprum，英文为 Copper）用"Cu"表示；氯（拉丁名为 Chlorum，英文为 Chlorine）用"Cl"；硫（拉丁名为 Sulphur，英文为 Sulfur）用"S"表示；硅（拉丁名为 Silicium，英文为 Silicon）用"Si"表示等。英文题目"Soil c and n contents affect microbial community structure"的正确表达应该是"Soil C and N contents

affect microbial community structure"；又如"Response of plants to elevated atmospheric CO_2"的正确表达应该是"Response of plants to elevated atmospheric CO_2"。❶

（3）专有名词首字母不大写。英语中的专有名词是特定的，包括特定的人名、地名、国名等。专有名词的首字母要大写，例如：China，Boston，Beijing South Railway Station，Dongting Lake，Chinese Academy of Sciences，American，International Energy Agency，China Daily，Dina，Smith 等。

（4）名词术语缩写不大写。名词术语缩写需要用大写，例如：APEC（Asia-Pacific Economic Cooperation，亚太经济合作组织），HTML（Hyper Text Markup Language，超文本标记语言），IDE（Integrated Development Environment，集成开发环境），EBA（Electronic Brake Assist，电子刹车辅助系统）等。英文题目"A rapid hiv infection in aferica"中"hiv"应该修改为"HIV"（"HIV"的英文全称为"Human Immunodeficiency Virus"，即"人类免疫缺陷病毒"）；"africa"应修改为"Africa"，即"非洲"。

（5）特殊表达没有大写。一些特殊表达的英文首字母需要大写；表示英文基本定理、模型的单词首字母要大写；pH 的"H"要大写等。例如："Mathematical expression of the first law of thermodynamics"，此处"the first law of thermodynamics"的正确表达应该为"the First Law of Thermodynamics"。英文题目"Determination of electrical capacitance of plant root system based on the dalton model"中的"the dalton model"应为"the Dalton Model"；英文题目"Spatiotemporal change of soil ph in China in past ten years"中的"ph"应为"pH"。

2.3.4 标题常见语法错误

（1）混淆词性。对一些意思比较接近的单词，不了解其词性，导致使用错误。例如，英文题目"A closed-form displacement affect of beam type structures to moving line loads"（移动线荷载作用下梁式结构的位移影响力闭合解），混淆了"affect"和"effect"的区别。两个英文单词都有"影响"的意思，但是前者是动词，后者是名词，正确的表达应该是"A closed-form displacement effect of beam type structures to moving line loads"。

（2）形近词分不清，导致拼写错误。例如："implicit"和"explicit"，"assess"和"access"，"beam"和"bean"，"clash"和"crush"，"adopt"和"adapt"，"quite"和"quite"，"angel"和"angle"，"principal"和"principle"，"abroad"和"aboard"，"contact"和"contract"等。

（3）冠词和介词使用错误。在英文中，冠词位于名词之前，是用于说明名词所指的人或物的虚词，其中"a"和"an"是不定冠词，"the"是定冠词。常见的误用之一是该使用冠词时没有使用，或不该使用冠词时却使用了；其二是该使用定冠词时使用了不定冠词，而该使用不定冠词时却又使用了定冠词，特别是一些固定表达中，不可以随意省略或添加，例如："in practice"不能写作"in the practice"，"on the other hand"不能写作"on other hand"等。介词也是英文题名中使用较多的一类虚词，有些介词跟动词、形容词和名词都有特定的搭配，需要按照规则使用。如"similar to"误用为"similar with"，"different from"误用为"different with"。

---

❶ 蔡卓平，段舜山，骆育敏. 科技论文英文题目的常见问题分析 [J]. 韶关学院学报（自然科学），2016，37（2）：67~70.

2.3.5 关键词的组成类型

（1）无特殊检索意义、不能表征文章所属学科专用概念的词不能独立作关键词使用，这类词 SCI 称为半禁用词（Semi-stop list），如：research, study, analysis, report, regulation, approach, theory, concept, idea, experimentation, phenomena 等。但半禁用词可以和半禁用词或实意词组合成关键词，如：理论研究、试验分析等。

  1）正：kinetic study
    误：kinetics
  2）正：decomposition mechanism
    误：thermal decomposition
  3）正：chemical reaction kinetics
    误：chemical reaction, kinetics

（2）无实质意义的词不能作关键词，这类词 SCI 称为全禁用词（Full-stop list）。如：冠词 articles（a/an, the），介词 prepositions（at, in, on, to, above, over, under, below…），连词 conjunctions（and, as well, but, while, yet…），代词 pronouns（I, me, mine, myself, who, this…），副词 adverbs，形容词 adjectives，感叹词 interjections，某些动词（连系动词、情感动词、助动词）some verbs（link verbs, modal verbs, auxiliary verb）等。

（3）尽量选用《汉语主题词表》中的主题词。选用有使用价值和一定的使用频率、能作为主题汇集一定量文献或具有叙词组配功能的名词术语。

  1）表示具有事物名称的名词术语，如：水泥、砂浆和混凝土（cement, mortar and concrete），建筑陶瓷（architectural pottery），建筑青铜合金（architectural bronze），汽车（automobile），变压器（transformer），反应堆（reactor），坐标仪（coordinatograph）等。

  2）表示事物的状态或现象的名词术语，如：风荷载（wind load），相对湿度（relative humidity），冻胀力（frost heave force），强度（strength/intensity），失真（distortion），土壤熟化（soil-ripening），船舶过载（marine overload）等。

  3）表示科学分类的名词术语，如：数学（mathematics），物理学（physics），中医学（traditional Chinese medicine（TCM）），电子学（electronics），建筑工程（constructional engineering），水利工程（hydraulic engineering）等。

  4）表示研究方法、技术方法的名词术语，如：分析化学（analytical chemistry），有限元法（finite element method），结构功能法（structure-function approach），力学性能实验（mechanical property test）等。

2.3.6 关键词的特殊选择

在下列情况下可选用自由词充当关键词：
（1）主题词表中明显漏选的主题词。
（2）表达新学科、新理论、新技术、新材料等新出现的概念。
（3）词表中未收录的地区、人物、文献、产品等名称及重要数据名称。
（4）某些概念采用组配，结果出现多义时，被标引概念也可用自由词。

2.4 范例阅读与分析

(1) 一种模糊省劲网络控制器与动态辨识器组成的控制系统

原：Control system of a fuzzy neural controller and its dynamical identifier

改：Control system based on a fuzzy neural controller and a dynamical identifier

原例中介词"of"表示所属关系，而且 its 指代不清，修改替换后使标题的意义更准确。

(2) 陈列天线通道误差的盲校正

原：The blind calibration of array antennae

改：Blind calibration of array antennae

原例中的定冠词就可以省去，这样既节省篇幅而且对文题没有影响。

(3) 一种基于二元语义信息处理的软件质量综合评价方法

原：An evaluation method for software quality based on two-tuple linguistic information processing

改：Evaluation method for software quality based on two-tuple linguistic information processing

原例中的不定冠词 an 可以省去，但要注意的是在有些情况下，可保留冠词，以示强调，也使文题的意义更加准确，如下面的两个例子。

(4) 墙壁开洞爆破技术

Techniques for blasting through a hole in the wall

(5) D-S 理论在信息融合中的改进

Improvement of D-S theory in an information fusion system

(6) 基于可用度的飞机可修件要求分析

原：The analysis of aircraft repairable spares requirement based on availability

改：Demand analysis of aircraft repairable spares based on availability

(7) 给予支持向量机的时间序列预测模型分析与应用

原：Time series forecasting model analysis and applications via support vector machines

改：Analysis and applications of time series forecasting model via support vector machines

修改后的内容以醒目的名词短语结构使整个标题更突出，符合标题准确、清楚、简洁的要求。

(8) CLUSTAL-W: improving the sensitivity of progressive multiple sequence alignment through sequence weighting, position-specific gap penalties and weight matrix choice

CLUSTAL-W（通过序列加权、位点特异性空位罚分和加权矩阵选择来提高渐进的多序列对比的灵敏度）。标题用 20 个词准确地表达了文章的多层意思：以最重要的词 CLUSTAL-W 作为标题的开头，紧接着解释 CLUSTAL-W 的目的是 improving the sensitivity of progressive multiple sequence alignment，达到该目的手段是 through sequence weighting, position-specific gap penalties and weight matrix choice。该标题重点突出、准确清楚，但似欠简洁。

（9）Processing of X-ray diffraction data collected in oscillation mode（震荡模式中X射线衍射数据的分析方法）。该标题准确、简洁、清楚，用7个实词和2个虚词清晰地说明了文章的研究主题内容为"Processing"，对象是"X-ray diffraction data collected in oscillation mode"（资料来源：唐耀，2005）

（10）Auditory Perspectives of Different Types of Music（合适）

（11）Electromagnetic Fields Have Harmful Effects on Humans（不合适）

改：Harmful Effects of Electromagnetic Fields on Humans

（12）How to Use Water Resources for Irrigation in Semiarid Land（不合适）

改：Using Water Resources for Irrigation in Semiarid Land

（13）Water Quality Can Be Protected Through the Successful Integration of Research and Education（不合适）

改：Protecting Water Quality Through the Successful Integration of Research and Education

（14）The Single Community Concept: A Model for Adult Environmental Education（合适）

（15）Physics and Art: Conceptual Linkages Can Be Uncovered（不合适）

改：Physics and Art: Uncovering Conceptual Linkages

（16）Diamond Is Used for Electronic Devices（不合适）

改：Use of Diamond for Electric Devices

（17）Yellow Fever's Effect on Transportation and Commerce（合适）

（18）The Nature of Student Science Project is compared with Educational Goals for Science（不合适）

改：The Nature of Student Science Projects in Comparison to Educational Goals for Science

（19）A Qualitative / Quantitative analysis of the Administrative Management Institute at Cornell University（合适）

（20）The Americans With Disability Act and Its Applicability to the Mentally Ill, Human Immune-Deficiency Virus and Acquired Immune Deficiency Syndrome Populations: A Statistical Analysis（不合适）

改：The Americans With Disability Act and Its Applicability to the Mentally Ill, Human Immune-Deficiency Virus and AIDS Populations: A Statistical Analysis

练习题

Task 1

Read the following titles and consider the key terms.

1. Cyclic seismic testing of composite concrete-filled U-shaped steel beam to H-shaped column connections.

2. Effects of cast-in-place concrete topping on flexural response of precast concrete hollow-core slabs.

3. Contribution of perforated steel ribs to load-carrying capacities of steel and concrete composite slabs under negative bending.

4. Who interacts on the Web? The intersections of user's personality and social media use.

5. A review of solar collectors and thermal energy storage in solar thermal applications.

6. The relationship between cell phone use, academic performance, anxiety, and satisfaction with life in

college students.

7. Experimental study on mechanical performance of checkered steel-encased concrete composite beam.

8. Behavior of a new shear connector for U-shaped steel-concrete hybrid beams.

Task 2

Fill in the blanks with appropriate prepositions.

1. Changes _____ the global value of ecosystem services.
2. Recent advances _____ graphene based polymer composites.
3. Relying _____ combined business of coal and electric.
4. Early identification systems _____ emerging foodborne hazards.
5. E-waste: An assessment _____ global production and environmental impacts.
6. Enlarging and excavating construction technology _____ vertical shaft.
7. Reducing enterprise cost and charge _____ strengthening management process control.
8. Analysis _____ hydraulic support in reliability.

Task 3

Read the following article and consider the appropriate title and key words.

Abstract: Nonlinear finite element analysis of the cast steel spherical joint is performed by using ANSYS program. The deformation process, the failure mode and the effects of parameters on bearing capacity for the joint are analyzed. The results show that the outer diameter of sphere D, wall thickness of sphere t, external diameter of cast steel tube d, radius of outer fillet r which is located in the connection of sphere to cast steel tube, material strength f have remarkable effects on the bearing capacity of the joint. The results also show that with the increasing of parameter t, d, r, and f, the bearing capacity increases, and with the increasing of parameter D, the bearing capacity decreases.

Introduction

Cast steel spherical joint is a more used way of cast steel joint at present, which includes the hollow sphere and tubular members molded by integral casting. Particularly in recent years, with the increasing of span of the structure and unusual geometry of joint, cast steel spherical joint has been paid more and more attention by the engineers for its serviceability. Despite many similarities on the shape between cast steel spherical joint and welded hollow spherical joint, bearing capacity and mechanical performance of these two are very different because of the differences of joint structures and manufacturing techniques. Nowadays, there are few studies on cast steel spherical joint, nor corresponding standards to guideline the design and reference books to introduce the calculating method. To evaluate the safety of the cast steel spherical joint for the specific structure, finite element analysis and experiments are used by which the joint stress and strain distribution can be obtained. However, this method brings many inconveniences for the design of cast steel spherical joint, which makes the joint not be applied widely. This paper is devoted to the study of the behavior of deformation, the failure mode and the effects of parameters on the bearing capacity for the cast steel spherical joint by program ANSYS. The results can provides reference for setting up the formula for bearing capacity of cast steel spherical joint.

The finite element model of cast steel spherical joint

As a result of being connected to multiple members, cast steel spherical joint in practice is subjected to complex three-dimension loads. Because it has similarities with welded hollow spherical joint on shape, that is they both have a closed spherical structure, therefore, they are subjected to multiple ring loads. What has been studied on welded hollow spherical joint indicates that the properties under 3-D load are similar with under one-way load and bearing capacity is mainly determined under one-way load. Hence, according to the research method of failure mechanism for welded hollow spherical joint, nonlinear finite element analysis of cast steel spherical joint with different parameters under one-way load are performed by program ANSYS.

As the shape of the joint and the distribution of loads are axisymmetric about central axis, when doing nonlinear analysis, the joint is simplified to a axisymmetric model to improve the speed of calculation. Besides, by comprehensively considering elements such as mesh scale, boundary condition, material characteristics and etc, a more reliable engineering accuracy is expected to be achieved for the calculation of bearing capacity. Four nodes plane element (plane 42) with axial symmetry, plasticity, large deformation, and large strain is selected. Each node has two degrees of freedom, that is the translations along X, Y. The material of cast steel is assumed to be perfectly elastic-plastic, the modulus of elasticity is $2.06 \times 10^5 \text{N/mm}^2$ and yield strength is 230 N/mm^2. We take Von Mises yield criterion and the corresponding flow rule. Because the wall of the cast steel spherical joint is more thick, for the purpose of improving calculation accuracy, the size of major element is $10 \sim 20\text{mm}$ and the length of element for fillet is reduced to half of the original size. The translations along the X, Y of the end of cast steel tube and symmetry of the joint are restrained. In order to describe nonlinear properties of load-displacement curve, the loads of the joint are applied by controlling the values of displacement. The freedoms corresponding to the vertical displacements of the end of cast steel tube are coupled, then the displacements are applied at the main freedom. In elastic stage, the values of displacement which are applied for every load step is 1mm, whereas the corresponding load step is divided to 10 load sub-step. When the material does not behave truly elastically, to make sure the results of nonlinear analysis matching load response, the values of applied displacement are similar to that in elastic stage, but the load step is divided to 20 load sub-step.

The behavior of deformation and the failure mode

The results of nonlinear finite element analysis shows the stress state is complex for outside fillets located at the connection of cast steel tube to sphere, and the stress reaches the maximum value here. On the condition that the cast steel tube is with reliable bearing capacity, when the joint is subjected to tensile force, the stress of the fillets reaches the yield strength of the material firstly and with the increasing of load, spherical face located near the fillets is pulled up, and the wall thickness of sphere becomes small gradually. Obvious plastic deformations and necking occur at the section where the fillet is tangential to the spherical face, and rupture occurs at last.

In view of the above analysis on the progress of deformation and failure mode of cast steel spherical joint, it can be seen when all the material of the sphere which is located near the outside fillet behave plastically, obvious plastic deformations occur, then the joint is insufficient to carry the loads and damaged, and the corresponding value of load is the critic load of joint for finite element analysis. Because of the wall thickness of the sphere and cast steel tube being large and the ratio of spherical diameter to wall thickness being less than 35, the failure of the cast steel spherical joint subjected to compressive force is not caused by buckling but insufficient strength. Consequently the above properties obtained by studying on the cast steel spherical joint under tension are suit for the joint under compression too. Therefore, the load converts from tensile force to compressive force for cast steel spherical joint, which will only lead to a change in the sign of stress and strain, whereas the distribution of stress and expansion progress of plastic area remains the same. According to the above definition, the influencing factors on bearing capacity of cast steel spherical joint subjected to tensile force are the same with those of joint subjected to compressive force.

Finite element parametric analysis of cast steel spherical joint

The results of elastic-plastic finite element analysis indicate that when the quality of material, casting technique and construction quality are assured, the main influencing factors on the bearing capacity of cast steel spherical joint include the outer diameter of hollow sphere D, wall thickness of hollow sphere t, external diameter of cast steel tube d, radius of outer fillet r which is located in the connection of the sphere to cast steel tube, and material strength f. Though the thickness of cast steel tube has no effect on bearing capacity, in order to avoid the cast tube being damaged firstly when the loads are applied, the value of ultimate load for cast steel tube should be calculated. The results of cast steel spherical joint with different size of inside fillet prove the angle of the inside fillet has almost no effect on bearing capacity. However, there are casting fillets in the inside of the real cast steel joint so as to meet the requirement of casting technique. Therefore, in order to match the real engineering condition we should take inner fillet into account for the finite element model, whose radius is the same with that of the outer fillet. In order to better know the influence rule of the parameters on bearing capacity, the diagrams of bearing capacity- parameter are obtained by plotting parameter as an abscissa and bearing capacity as an ordinate.

Bearing capacity of cast steel spherical joint is tightly related to strength f. With the increasing of strength f, bearing capacity strengthens correspondingly, and it shows there exists a direct proportion between them, which in turn proves the assertion of the failure mode for cast steel spherical joint.

The effect of spherical diameter on bearing capacity

Bearing capacity of cast steel spherical joint tends to decrease along with the increase of the spherical diameter, but with a small change on the value of bearing capacity. The bigger diameter gets, the more uneven distribution of stress for the connection of sphere to cast steel tube becomes, and the more serious stress concentration gets. Consequently, bearing capacity decreases.

The effect of wall thickness of sphere on bearing capacity

With the wall thickness of sphere becoming larger, bearing capacity increases remarkably, and the relation between them is directproportion. As the position of failure occurs at the section located at the outer fillets in the connection of the cast steel tube to the sphere, with the increase of the wall thickness, the area of failure surface extends accordingly, which leads to a decrease in the stress in the cross section but an increase in bearing capacity.

The effect of diameter of cast steel tube on bearing capacity

The outer diameter of cast steel tube affects bearing capacity greatly. The bigger the diameter of cast steel tube is, the higher the bearing capacity will be. The area of failure surface extends following the increasing of diameter, which reduces the stress in the cross section and improves the bearing capacity.

The effect of radius of fillet on bearing capacity

The radius of fillet has some influences on bearing capacity. With the increasing of the radius, bearing capacity goes up accordingly. The fillet not only reduces the peak value of stress, but also improve the stress concentration, and with the growing of radius, the area of failure surface increases, which causes areduced stress in the cross section whereas an increased bearing capacity on joint.

Conclusions

The results of nonlinear finite element analysis for cast steel spherical joint by ANSYS program indicates that the position of failure is located in the connection of the cast steel tube to sphere. When all material of the sphere which is located at the outer fillet behave plastically, obvious plastic deformations occur and the joint is damaged.

Because the wall thicknesses of sphere are larger, which has exceeded that of the thin shell, the failure mode of joint subjected to compressive force is the same with that of joint subjected to tensile force, and they all belong to the failure due to insufficient strength.

The effects of parameters of cast steel spherical joint on bearing capacity are analyzed. With the increasing of spherical diameter D, the capacity decreases, and with the increasing of the thickness t, diameter of cast steel tube d, fillet radius r, strength f, the bearing capacity increases.

Task 4

Read the following titles, abstracts and consider the appropriate key words.

1. Title: Analysis of parameter influence on seismic behavior of concrete filled square steel tubular columns

Abstract: In order to study the influences of several parameters such as the steel ratio, slenderness ratio and compression ratio on seismic behaviors of concrete filled square steel tubular (CFST) columns, a finite element analysis model was proposed. Compared with experimental hysteretic curves, it is shown that computational results are acceptable, indicating that the model proposed could be applied to the seismic analysis of CFST columns. Based on the model, CFST columns with different parameters were simulated. The results show that the load-displacement hysteretic loops of the CFST columns are plump in shapes, and have no obvious pinching phenomenon. Furthermore, the slenderness ratio and the compression ratio have great impact on the deformation capacity of CFST columns. In addition, the steel ratio and the slenderness ratio are the most influential factors on the bearing capacity of CFST columns. Among the three parameters, the slenderness ratio is the most significant factor on stiffness degradation of CFST columns. Within a certain range, the CFST column with high steel ratio and compression ratio and a low slenderness ratio has a better energy dissipating capacity. (资料来源：王铁成，2013)

2. Title: Seismic Behavior of Diaphragm-Through Connections of Concrete-Filled Square Steel Tubular Columns and H-Shaped Steel Beams

Abstract: Based on the introductions of a type of diaphragm-through connection between concrete-filled square steel tubular columns (CFSSTCs) and H-shaped steel beams, a finite element model of the connection is developed and used to investigate the seismic behavior of the connection. The results of the finite element model are validated by a set of cyclic loading tests. The cyclic loading tests and the finite element analyses indicate that the failure mode of the suggested connections is plastic hinge at the beam with inelastic rotation angle exceeding 0.04 rad. The suggested connections have sufficient strength, plastic deformation and energy dissipation capacity to be used in composite moment frames as beam-to-column rigid connections. （资料来源：荣彬，2013）

3. Title：Risk assessment in non-standard forms of civil engineering consulting services

Abstract：Although a large body of research exists on risk assessment in civil engineering projects and of owners, con-tractors, concessionaires and financiers of such projects, there is a lacuna in such research on engineering consultants, particularly those associated with non-standard forms of consulting services. This paper seeks to explore the genesis of the underlying risks in non-standard forms of engineering consulting services, systematically classify the risks, and develop a Risk Breakdown Structure and a Generic Framework for efficient assessment of these risks, which is a prerequisite for sound risk management in the engineering consulting industry. The research adopts a mixed method approach, synthesising exploratory type multiple-case studies and questionnaire surveys, carried out in 14 engineering consulting firms having extensive experience in the delivery of non-standard consulting services. This paper provides empirical insights of the genetic make up of risks associated with non-standard forms of consulting services. Such risks are found to be predominantly linked to design office based activities that underline the importance of design function in engineering consulting practice. Loss of reputation and/or goodwill is rated as the most severe potential impact on consultants. Proposed Risk Assessment Framework provides the engineering consulting industry with a functional tool for efficient risk management. （资料来源：WM Sarath C PIYADASA，2014）

参考文献

[1] 张超英，等. 科技论文英文题名制作的技术规范 [J]. 编辑学报，2006（12）：14~16.

[2] 李平，曹雁. 科技期刊论文英文标题抽象名词短语结构分析与应用 [J]. 中国科技期刊研究，2012，23（2）：322~324.

[3] 任胜利. 科技论文英文题名的撰写 [J]. 中国科技期刊研究，2003，14（5）：567~570.

[4] 唐耀. 谈谈科技论文的英文标题. 见：第5届中国科技期刊青年编辑学术研讨会论文集 [C] //河南：中国学术期刊（光盘版）电子杂志社，2005：29~31.

[5] 朱彩萍. 学术论文中关键词的规范 [J]. 图书情报，2005（4）：51~53.

[6] 崔鲸涛. 科技论文英文标题制作的技术规范 [J]. 应用写作，2003（7）：41~42.

[7] 任芬梅. 科技论文写作中英文题名的规范表达 [J]. 中北大学学报（社会科学版），2007，23（1）：98~100.

[8] 杜兴梅. 学术论文摘要与关键词的写作及其格式规范 [J]. 韩山师范学院学报，2008，29（2）：82~87.

[9] 郭庆健，孙守增，芮海田. 科技论文的题名、摘要和关键词的写作 [J]. 长安大学学报（自然科学版），2005，25（1）：108~110.

[10] 陆艾五. 科技论文的中英文标题写作 [J]. 安徽农学院学报，1990，10（2）：131~132.

[11] 蔡卓平，段舜山，骆育敏. 科技论文英文题目的常见问题分析 [J]. 韶关学院学报（自然科学），2016，37（2）：67~70.

[12] Lin Y, et al. Finite element parametric analysis on bearing capacity of cast steel spherical joint [J]. Applied Mechanics and Materials，2011，94~96：365~368.

[13] 闫莉，孙洪丽，程文华. 英语研究论文写作 [M]. 上海：上海交通大学出版社，2017：4.

[14] 张荔. 学术英语交流——写作与演讲 [M]. 上海：上海交通大学出版社，2017：11~23.

[15] 刘园丽. 大学学术英语写作研究［M］. 北京：中国水利水电出版社，2019：37~42.
[16] B7713-87 科学技术报告、学位论文和学术论文的编写格式.
[17] CAJ-CDB/T1-1998，中国学术期刊（光盘版）检索与评价数据规范.
[18] https：//jingyan. baidu. com/article/d45ad148c908e069552b8005. html.
[19] https：//wenku. baidu. com/view/ddc194600b1c59eef8c7b4f5. html.
[20] http：//www. docin. com/p-37953448. html.

3 摘 要

3.1 概 述

3.1.1 摘要的定义

论文摘要又称概要、内容提要,是对论文的内容不加注释和评论的简短陈述,《MLA 格式指南及学术出版准则》要求摘要最长不超过 350 字,其基本要素包括研究目的、方法、结果和结论。具体地讲就是研究工作的主要对象和范围,采用的手段和方法,得出的结果和重要的结论,有时也包括具有情报价值的其他重要的信息,重点是结论,是一篇具有独立性和完整性的短文。

3.1.2 摘要的分类

论文摘要一般分为描述性摘要和信息性摘要。描述性摘要也称为指示性摘要,简要介绍论题和研究目的或者论文框架。信息性摘要包括主题、背景回顾、研究数据方法、研究结论和研究用途。

根据内容的不同,摘要可分为以下三类:报道性摘要、指示性摘要和报道指示性摘要。

(1) 报道性摘要:常称作信息性摘要或资料性摘要,其特点是全面、简要地概括论文的目的、方法、主要数据和结论。通常,这种摘要可以部分取代阅读全文。科技论文如果没有创新内容,如果没有经得起检验的与众不同的方法或结论,是不会引起读者的阅读兴趣的;所以建议学术性期刊(或论文集)多选用报道性摘要,用比其他类摘要字数稍多的篇幅,向读者介绍论文的主要内容。篇幅以 300 字左右为宜。

(2) 指示性摘要:常称为说明性摘要、描述性摘要或论点摘要,一般只用 2~3 句话概括论文的主题,而不涉及论据和结论,多用于综述、会议报告等。该类摘要可用于帮助潜在的读者来决定是否需要阅读全文。篇幅以 100 字左右为宜。

(3) 报道指示性摘要:以报道性摘要的形式表述一次文献中的信息价值较高的部分,以指示性摘要的形式表述其余部分。将上述的两种摘要的内容有机结合便是报道-指示性摘要。篇幅以 100~200 字为宜。

以上 3 种摘要分类形式都可供作者选用。一般地说,向学术性期刊投稿,应选用报道性摘要形式;只有创新内容较少的论文,其摘要可写成报道-指示性或指示性摘要。论文发表的最终目的是要被人利用。如果摘要写得不好,论文进入文摘杂志、检索数据库,被人阅读、引用的机会就会少得多,甚至丧失机会。一篇论文价值很高,创新内容很多,若写成指示性摘要,可能就会失去较多的读者。

3.1.3 摘要的功能

论文摘要如同人体中的心脏，起着核心的作用。作为论文中最重要的部分，摘要通常具有以下功能：

（1）摘要担负着吸引读者和将文章的主要内容介绍给读者的任务，可以为科技情报文献检索数据库的建设和维护提供方便。因此，摘要应具有独立性和自含性，即不阅读报告、论文的全文，就能获得必要的信息。让读者尽快了解论文的主要内容，以补充题名的不足。论文摘要的质量高低，直接影响着论文被检索率和被引的频次。

这里要特别强调英文摘要的完整性，即英文摘要所提供的信息必须是完整的。因为论文是用中文写作的，中文读者在看了中文摘要后，不详之处还可以从论文全文中获得全面、详细的信息，但英文读者（论文的编辑）如果看不懂中文，英文摘要就成了他唯一的信息源。即使读者看不懂中文，也可以通过英文摘要对论文的主要目的，解决问题的主要方法、过程，及主要的结果、结论和文章的创新、独到之处，有一个较为完整的了解。

因此，在写英文摘要时，不能简单地把论文前面的中文摘要翻译成英文，而要保证英文摘要内容的独立完整性。

（2）摘要中的定量分析，能够体现研究的科学性。注重定量分析是科学研究的重要特征之一。这一点也应该体现在中英文摘要的写作中。因此，在写作摘要时，要避免过于笼统的、空洞无物的一般论述和结论。要尽量利用论文中的最具体准确的语言来阐述作者的方法、过程、结果和结论，这样既可以给读者一个清晰的思路，又可以使论述言之有物、有根有据，使读者对作者的研究工作有一个清晰、全面的认识。

（3）摘要能够体现论文和研究的创新性。写论文时要明确地突出自己的贡献，突出自己的创新、独到之处。读者在阅读论文时也总是特别关注论文有什么创新独到之处，否则就认为论文是不值得读的。另外，由于东、西方文化传统存在很大的差别，我国长期以来的传统教育有些过分强调知识分子要"谦虚谨慎、戒骄戒躁"，因此我国学者在写作论文时，一般不注重（或不敢）突出表现自己所做的贡献。

由于中、英文摘要的读者对象不同，鉴于上述两方面的因素，笔者认为论文中的中、英文摘要不必强求完全一致。

3.1.4 摘要的基本要素

摘要的四要素包括论文研究目的、方法、结果和结论。

（1）目的：论文研究的范围、目的和重要性。目的部分应简要说明研究的目的，说明提出问题的缘由，表明研究的范围及重要性。

（2）方法：论文使用了哪些研究方法。方法部分应说明研究课题的基本设计，使用了什么材料和方法，如何分组对照，研究范围以及精确程度，数据是如何取得的以及经过何种统计学方法处理。

（3）结果：陈述论文的研究所取得的成果。结果部分要列出研究的主要结果和数据，有什么新发现，说明其价值及局限，叙述要具体、准确，并需给出结果的可信值和统计学显著性检验的确切值。

（4）结论：论文的研究所得出的重要结论及主要观点。结论部分应简要说明、论证取得的正确观点及其理论价值或应用价值，是否值得推荐或推广等。

3.2 摘要的写作范式

3.2.1 研究主题的写作范式

摘要的研究主题句一般描述研究对象、目的、解决的问题，通常采用以下范式（斜杠为并列的可替换内容）：This article/analysis/essay/investigation/paper/research/study/survey/thesis/work addresses/analyzes/argues/deems/holds/concerns/covers/includes/deals with/touches on/is about…

The author demonstrates/discusses/describes/elaborates on/expounds/emphasizes/explains/explores/probes into/expresses/focuses on/formulates/introduces/presents/investigates/monitors/provides/reports/reveals/suggests/shows/exhibits/studies/summarizes…或使用被动语态或添加主语，例如：…is studied in this paper/ is the subject of this investigation.

3.2.2 研究背景的写作范式

研究背景需要综述国内外关于同类课题研究的现状，从而提出问题，阐述研究该课题的原因。研究背景包括理论背景和现实需要，综述国内外关于同类课题研究的现状：（1）前人研究的成果，所选题目到目前所研究到的状况，而你又对选题有何特别看法。（2）找出你想研究而前人还没有做的问题。（3）前人已做过，你认为做得不够（或有缺陷），提出完善的想法或措施。（4）前人已做过，你重做实验来验证。

研究背景（背景回顾）的写作范式有：

…have developed dramatically over the last century, lessons learned…

In the previous publication it was shown…

This paper presents the results of…

…surveys conducted by…

This paper provides a brief history of…

The paper reviews the context for…

研究背景（必要性和重要性）的写作范式有：

This analysis/article/essay/paper/thesis considers/deems it necessary to…

It is necessary to…

…create (s) new problems and call (s) for new…

…has (have) resulted in a need for…

…require (s) an effective…

…is an essential process to…

…play (s) important roles in…

…is an important means of…

…will be important.

研究背景（局限性）的写作范式有：

Previous surveys have addressed the question of…. The surveys reported in this paper are a follow-up study to the earlier surveys.

Although this process has successfully achieved this objective, it is still not completely understood…. There is limited information regarding…and no information could be found on the relationship between…and…. To understand…

The limitations of…are discussed.

……

has the potential to…. However, no comparative studies on…are available, and…has not been attempted.

……

is a major contributor to…. However it is much neglected in the research and planning activities of…

Despite the recent…of…

3.2.3 研究目的的写作范式

首先,要说明作者写此文的目的,或者本文要解决的主要问题。一篇好的英文摘要,一开头就应该把要写作的目的或要解决的主要问题非常明确地交代清楚。在突显本项目研究的必要性和重要性的同时,进一步详细说明本研究的具体研究目的和范围,使后续的讨论更具有针对性。用陈述目的的主题句开始。这种方法开门见山,直截了当,使读者一下就抓住文摘的中心。该项内容通常就是用一句话来完成,必须紧扣论文题目和主题。通常采用一般现在时或直接用动词不定式词组来表达。如:

This paper describes recent modeling and experimental studies of reverse combustion (RC) linking, aimed at understanding the propagation of dynamics of a RC front.

In what follows, this dissertation wants to…

The primary purposes/The main goals/The ultimate aims/The problems of interest/The intentions for this study/of this work/of this research/of this dissertation/of this paper/of this thesis…

研究目的要写得具体,要能吸引读者,要给读者留下深刻的印象。常用下列句型:

(1) This paper presents an approach to equipment reliability prediction based on the concept that failures of electronic equipment are ultimately due to chemical, mechanical and/or metallurgical processes.

(2) The discussion of…issue forms the major concern of the present study.

(3) This paper develops a theoretical framework to evaluate the benefits and costs of energy projects in oil-producing developing countries.

(4) The analysis/article/essay/paper/thesis aims to/attempts to/makes an attempt to/tries to/intends to/is intended to…

The chief/key/main/major/primary/principal/aim/attempt/goal/object/objective/proposal/purpose of this paper/research/work is to…

To/In order to assess/evaluate/compare/clarify/describe/determine/explore/examine/testify/check/identify/improve/investigate/study…

To this end…is (was) studied…

In this paper, we wish to…

(5) This paper (report, thesis, work, presentation, document, account, etc.) describes (reports, explains, outlines, summarizes, documents, evaluates, surveys, develops, investigates, discusses, focuses on, analyzes, etc.) the results (approach, role, framework, etc.) of…

(6) This report (article, paper, etc.) addresses (is concerned with, argues, specifies, covers, etc.) the following questions…

(7) Our goal has been to develop…

(8) This research project is devoted to…

(9) The objective (purpose, motivation, etc.) of this paper (report, program, etc.) is…

(10) This paper has three main objectives…

研究目的也可用名词短语前置的句型表示，把重点内容放在最显著的位置加以强调，这样的文体更加正式一些。例如：

(1) The first known measurement of the differential cross section for electron capture to the continuum (ECC) from atomic hydrogen is presented.

(2) Procedures for testing atmospheric transport and dispersion models for distances of several hundred to 1000km from sources of pollutants are reviewed.

为了避免句子过分的头重脚轻，使句子结构更加平衡，也常使用下列句式：

(1) An account is provided of…

(2) Detailed information is presented about…

(3) New results are presented of studies of the application of inorganic exchangers in the following fields…

3.2.4 研究方法的写作范式

研究方法是对研究过程的描述，其作用是明确告诉读者研究使用了什么样的方法，如文献研究法、实验研究法等。陈述研究实验中的对象、实验研究中曾使用的材料、设备、工艺、手段、程序等方面。该项内容包括较多，组织较难，一定要用较为简单的语言来表达该项内容。因该项目所表述的是已经采用的实验方法、路线和手段，故常用一般现在时。

研究方法（方法与原理）的写作范式有：

A…based…is used, which requires…

A…model is developed to represent…

A…model tested the …of…

A simple modification in the model changes the focus from …to…

A…system was used to…

…was (were) treated with…

…measurements were made…

…were separated into…groups based on…

…were randomly divided/ grouped into…groups.

Statistical methods…was used for…

…also incorporates a…

The new…system has been tested in…

The proposed…employs an information system to store…

The samples of…were collected by…

The test for…has been carried out…

We sampled the…of…

We experimented on the property of…

Using…(technique), we studied…

Using…, it was found that…

…was (were) measured using…

By performing a one-dimensional analysis on a…

With reference to…, …is obtained, …

Problem is solved through…method.

In order to examine the…, a questionnaire survey has been conducted with…

研究方法（数据测量与计算）的写作范式有：

The average…increases with…

The mean percent differences between…and…

that were observed indicated that…

and it was evident that…

The author has computed/ worked out…

In this paper, we measured…

Comparing…and…, it was found that…

The…range of…was measured by…

The…rate of…was calculated by means of…

The…was estimated…

3.2.5 研究结果的写作范式

研究结果是研究问题的答案及研究发现。揭示已经得出的研究结果，包括数据、效果、性能等方面，常采用一般过去时。

其写作范式有：

It is found that…

The…are found to be…

…is confirmed.

…demonstrate…

…showed…

Key findings of the study include…

3.2.6 研究结论的写作范式

在摘要的最后部分,作者要对全文作一总结或对以后的研究提出建议。研究结论是实验、观测结果和理论分析的逻辑发展,是整篇论文的总论点。结论部分主要证明研究引发的思考。根据研究结果提出问题、建议、预测,包括对结果的分析、比较、应用等方面。此项内容包括通过实验研究或总结,对于同行和阅读者有什么样的帮助、建议和启迪,所以常用一般现在时或情态动词。

研究结论(结论)的写作范式有:
(1) We thus conclude that…
(2) The result shows that…
(3) We conjecture that…
(4) Although a number of tests and comparison of the method have given satisfactory results, additional investigations to provide further justification and verification are required.
(5) The study of……indicates/reveals/shows/suggests that…
(6) The authors conclude that…

研究结论(启示)的写作范式有:
The approach can be used with…
Based on…, …can determine…
The model can be used for answering various "what if" questions that may…

3.3 范文阅读与分析

3.3.1 相关阅读:英文摘要的注意事项

(1) 英文摘要的时态。英文摘要时态的运用也以简练为宜,常用一般现在时、一般过去时,很少用现在完成时、过去完成时,进行时态和其他复合时态则基本不用。

一般现在时:一般用于说明研究目的、叙述研究内容、描述结果、得出结论、提出建议或讨论等。涉及公认事实、自然规律、永恒真理等,当然也要用一般现在时。例如:

To solve this problem, a control system based on LORA technology of low power and long distance wireless communication is proposed.❶(研究目的)

Through parameter optimization configuration and network camouflage, the disguised access point of the IOT terminal is established. And at the same time, the tracking area code is updated to make IOT terminals carry out updating their locations. Finally, the number of IOT terminals and their identity information are extracted, laying a foundation for big data collection and identification of illegal terminals subsequently in space of the IOT. Through experimental verification, the technology has a higher probability of detection and identification, and it has a certain

---

❶ 曹晟,王昌栋,邢建春,等. 基于 LORA 物联网的人防工程三防控制系统 [J]. 计算机与控制系统,2019,38(10):118~121.

application prospect. ❶（研究内容）

The results show that：（1）The workers have a high degree of cognition bias from severity of the accident and availability of accident risk, and a lower degree of cognition bias from likelihood of accident. （2）Hazard cognition of the workers mainly depends on empirical information and subjective feelings. Representativeness bias and availability heuristic, stereotype effect and anchoring effect, overconfidence and Dunning-Kruger effect, which is prone to have when recognizing likelihood of the accident, severity of the accident, and controllability of accident risk. （3）The mismatch between the empirical information and the actual situation, the excessive reliance on the subjective one-sided feeling of danger, and the negative impact of poor safety atmosphere account for hazard cognition bias of the workers. ❷（研究结果）

涉及公认事实、自然规律、永恒真理等，当然也要用一般现在时。

一般过去时：用于叙述过去某一时刻（时段）的发现、某一研究过程（实验、观察、调查、医疗等过程）。例如：

Remote sensing images of earthquake disaster areas were collected by UAV, and the problem of building structure deformation detection was transformed into a problem of coordinate measurement between components. Sample vector points in the collected remote sensing images were extracted and divided into different kinds of regions. On this basis, the images were clustered and segmented to obtain the characteristics of different types of building structures in post- earthquake images, thus identifying the earthquake disaster situation of different sample vector points. ❸

另外，需要特别注意的是，用一般过去时所描述的发现、现象和过程，往往是尚不能确认为自然规律、永恒真理的，而只是实验、调查当时如何如何；所描述的研究过程也明显带有过去时间的痕迹。

现在完成时和过去完成时：完成时尽量少用，但不是完全不用。其中，过去完成时可用来表示过去某一时间以前已经完成的事情，或在一个过去事情完成之前就已完成的另一过去行为。例如：

The advantages of prefabricated buildings have been demonstrated by lots of scholars.

Concrete has been studied for many years. Man has not yet learned to store the solar energy.

The microwave thermal utilization technology has promoted the application study of microwave absorbing materials.

（2）英文摘要的语态。采用何种语态，既要考虑摘要的特点，又要满足表达的流畅。由于一篇摘要一般很简短，尽量不要随便混用语态，更不要在一个句子里混用。

主动语态：有助于摘要的文字清晰、简洁及表达有力。例如：

---

❶ 郭悦，王红军. 基于中间人的物联网终端探测和识别［J］. 电讯技术，2019，59（10）：1197～1202.

❷ 韩豫，尹贞贞，刘嘉伦，等. 建筑工人的危险认知偏差特性及成因［J］. 土木工程与管理学报，2019，36（5）：56～67.

❸ 刘定操，尚展垒. 利用遥感图像对震损建筑结构变形检测的识别研究［J］. 地震工程学报，2019，41（5）：1380～1384.

The author systematically introduces the history and development of Indian culture.

The history and development of Indian culture are introduced systematically.

两相比较第一句要比第二句的语感更强、更简洁，信息更丰富、更明确。举例如下：

Using evolutionary game theory, the study constructs a three-dimension dynamic evolutionary game model about shared social responsibility between the energy-saving service enterprises and the existing construction owners under the supervision of the government in the three phases of energy-saving reform, the implementation stage and the operation stage. ❶

The test results show that the system has a stable performance, users could do timely intervention control, set operational parameters and collect environmental parameters, and the view of historical data and the online pay for server hosting could meet the practical application requirements. ❷

被动语态：以前强调多用被动语态，因为科技论文主要是说明研究事实经过，至于那件事是谁做的，无须一一明确说明。被动语态可以在语篇中强调主题，在指示性摘要中，为强调动作承受者，还是采用被动语态更好。而在报道性摘要中，有些情况下被动者无关紧要，也必须用被强调的事物做主语。举例如下：

However, CAT is far behind the development of CAD. Aiming at this problem, CAT is researched in this paper. New methods are put forward for judging positive or negative character of each composing loop and making it easier for loop searching, based on the position data of fits from API interfaces. A CAT plug-in integrated into SolidWorks is developed, to generate tolerance design function automatically and realize a higher efficiency of tolerance optimization design by simply interacting with 3D models. ❸

On basis of it, absorbing heat generation properties of carbon nanotube were explored under microwave radiation. ❹

（3）英文摘要的人称。英文摘要同中文摘要相比，英文摘要有其特殊性。一方面，英、汉语言自身存在较大的差异；另一方面，期刊所附的英文摘要基本上都是从中文摘要翻译过来的，这就涉及"对等"与"不对等"的问题，是否中英文要完全一致的问题。加之国家标准对摘要撰写的硬性约束，我国不少期刊的英文摘要也要求使用第三人称。因此，行文时最好不用第一人称，以方便文摘刊物的编辑刊用。例如：

Based on this, this paper first puts forward relevant concepts, explores the problems existing in computer network technology, and finally analyzes the application of artificial intelligence tech-

---

❶ 郭汉丁，张印贤，陈思敏．既有建筑节能改造市场主体各阶段社会责任共担演化机理［J］．土木工程与管理学报，2019，36（5）：25~32．

❷ 左志宇，谭洁，毛罕平，等．基于物联网的微型植物工厂智能监控系统设计［J］．农机化研究，2019（11）：74~79．

❸ 赵方舟，罗大兵，陈达，等．基于SolidWorks的计算机辅助公差优化设计研究［J］．机械设计与制造，2019（10）：15~19．

❹ 李哲，王文龙，张梦，等．碳纳米管材料低频电磁参数及吸波产热特性［J］化工学报，2019，70（S1）：28~34．

nology in computer network technology. ❶

This paper investigates the influence of subway induced vibration on a proposed residential office building. It performed filed measurements lasting 24 hours at this building site. ❷

3.3.2 范文分析

范文 1：

纳米粒子对熔盐复合蓄热材料热物性的影响

摘要：为了得到 SiO_2 纳米粒子含量对 $SiO_2/NaNO_3$-KNO_3/EG 复合蓄热材料比热容和热导率的影响，通过机械分散法，采用 $NaNO_3$-KNO_3 和不同质量分数（0.1%，0.5%，1%，2%，3%）的 SiO_2 纳米粒子所形成的熔盐纳米材料作为蓄热材料，膨胀石墨（EG）作为基体材料，制备出纳米 $SiO_2/NaNO_3$-KNO_3/EG 复合材料。对复合材料的比热容和热导率进行了测量，同时用扫描电镜对其微观结构特征进行了分析。结果表明，SiO_2 纳米粒子的质量分数为 1% 时，复合材料的平均比热容和热导率分别为 3.92J/(g·K) 和 8.47W/(m·K)，与其他纳米 SiO_2 添加比例相比，其比热容和热导率分别提高了 1.37~2.17 倍和 1.7~3.2 倍。这是由于复合材料表面会形成高密度的网状结构，这种具有较大比表面积和高表面能的特殊纳米结构可以提高复合材料的比热容和热导率。

Abstract：In order to study the effects of SiO_2 nanoparticles content on the specific heat capacity and thermal conductivity of nano-$SiO_2/NaNO_3$-KNO_3/EG composite heat storage materials, a series of nano-$SiO_2/NaNO_3$-KNO_3/EG composites were prepared by mechanical dispersion method. $NaNO_3$-KNO_3 and SiO_2 nanoparticles with different mass fractions (0.1%, 0.5%, 1%, 2%, 3%) were used as heat storage materials and expanded graphite (EG) was used as matrix material. Then the specific heat and the thermal diffusivity of composite heat storage materials were measured, and the microstructural characteristics were analyzed by scanning electron microscopy (SEM). The results show that adding 1% of SiO_2 nanoparticles to the composite can significantly affect its average specific heat capacity and thermal conductivity, with a measured value of 3.92J/(g·K) and 8.47W/(m·K), respectively, which are 1.37~2.17 times and 1.7~3.2 times higher than that of the other similar composites. This is owing to its high density network nanostructure with the large specific surface area and high surface energy which can improve the specific heat capacity and thermal conductivity.

范文分析：

研究目的和方法：In order to study the effects of SiO_2 nanoparticles content on the specific

---

❶ 张文娟，张海涛. 大数据时代人工智能在计算机网络技术中的运用 [J]. 网络信息工程，2019 (9)：73~74.

❷ 高广运，耿建龙，毕俊伟，等. 地铁环境振动对建筑场地影响实测分析 [J]. 工程地质学报，2019，27（5）：1116~1121.

heat capacity and thermal conductivity of nano−SiO_2/$NaNO_3$−KNO_3/EG composite heat storage materials, a series of nano−SiO_2/$NaNO_3$−KNO_3/ EG composites were prepared by mechanical dispersion method.

研究方法：$NaNO_3$−KNO_3 and SiO_2 nanoparticles with different mass fractions (0.1%, 0.5%, 1%, 2%, 3%) were used as heat storage materials and expanded graphite (EG) was used as matrix material. Then the specific heat and the thermal diffusivity of composite heat storage materials were measured, and the microstructural characteristics were analyzed by scanning electron microscopy (SEM).

研究结果：The results show that adding 1% of SiO_2 nanoparticles to the composite can significantly affect its average specific heat capacity and thermal conductivity, with a measured value of 3.92J/(g·K) and 8.47W/(m·K), respectively, which are 1.37~2.17 times and 1.7~3.2 times higher than that of the other similar composites.

研究结论：This is owing to its high density network nanostructure with the large specific surface area and high surface energy which can improve the specific heat capacity and thermal conductivity.

本篇摘要比较简短，没有涉及研究背景。本摘要节选自《化工学报》2019 年第 70 卷第 S1 期，论文作者：于强，鹿院卫，张晓盼，吴玉庭。

范文 2：

建筑工人的危险认知偏差特性及成因

摘要：为更有效地纠正建筑工人危险认知偏差，预防不安全行为发生，以认知心理学为基础，先采用心理测量方式获取工人的危险认知结果，再以对比方式反映偏差程度，最后对偏差成因进行探索分析。研究表明：(1) 工人对事故后果严重性和风险可控性的认知偏差程度较高，对事故发生可能性的认知偏差程度较低；(2) 工人进行危险认知时主要依赖经验信息和主观感受，对事故发生可能性进行认知时容易发生代表性偏差和易得性偏差，对事故后果严重性进行认知时容易发生定型效应和锚定效应，对事故风险可控性进行认知时易出现过度自信和达克效应；(3) 认知所需的经验信息与实际状况不匹配，认知中过于依赖对危险的主观片面感受，不佳安全氛围的负面影响等是造成工人危险认知偏差的主要原因。

Characteristics and Causes of Hazard Cognitive Bias of Construction Workers

Abstract: To correct hazard cognition bias of construction workers more effectively, and prevent unsafe behavior occurring, first, hazard cognition results of the workers were obtained by psychological measurement. Then the degree of bias was reflected by contrast. At last the causes of bias were explored and analyzed. All above are based on cognitive psychology. The results show that: (1) The workers have a high degree of cognition bias from severity of the accident and availability of accident risk, and a lower degree of cognition bias from likelihood of accident. (2) Hazard cognition of the workers mainly depends on empirical information and subjective feelings. Representativeness bias and availability heuristic, stereotype effect and anchoring effect, overconfidence and Dunning-Kruger effect, which is prone to have when recognizing likelihood of the accident, severity of the accident, and controllability of accident risk. (3) The mismatch between the empirical information and the actual situation, the excessive reliance on the subjective one-sided feeling of danger, and the negative impact of poor safety atmosphere account for hazard cognition bias of the workers.

范文分析：

研究目的：To correct hazard cognition bias of construction workers more effectively, and prevent unsafe behavior occurring.

研究方法：First, hazard cognition results of the workers were obtained by psychological measurement. Then the degree of bias was reflected by contrast. At last the causes of bias were explored and analyzed.

研究结论：The results show that: (1) The workers have a high degree of cognition bias from severity of the accident and availability of accident risk, and a lower degree of cognition bias from likelihood of accident. (2) Hazard cognition of the workers mainly depends on empirical information and subjective feelings. Representativeness bias and availability heuristic, stereotype effect and anchoring effect, overconfidence and Dunning-Kruger effect, which is prone to have when recognizing likelihood of the accident, severity of the accident, and controllability of accident risk. (3) The mismatch between the empirical information and the actual situation, the excessive reliance on the subjective one-sided feeling of danger, and the negative impact of poor safety atmosphere account for hazard cognition bias of the workers.

本篇摘要比较简短，没有涉及研究背景和研究结果等。研究目的和研究方法出现在同一句话里。本摘要节选自《土木工程与管理学报》第 36 卷第 5 期（2019 年 9 月），论文作者：韩豫，尹贞贞，刘嘉伦，冯志达，金若愚。

范文 3：

内廊式建筑火灾外部烟气蔓延规律

摘要：为了探究内廊式建筑火灾外部烟气蔓延规律，以某内廊式建筑为研究背景，通过数值模拟研究不同火源特征参数下外部烟气的形成机理及蔓延规律。研究发现：火灾前期，室内烟气会发生溢出形成外部烟气，在临近上层走廊重新进入室内，导致温度在短时间内显著增大，随后外部烟气逐渐减少，温度逐渐降低并保持稳定；当火灾规模较小时，火源位置变化对外部烟气蔓延的影响不显著；火灾规模较大时，火源位置存在明显的危险区段：火源处于该区段中，外部烟气的影响程度显著增大，在区段外影响程度较低，危险区段为距离通风口 4.5~6.5m；临界风速可以作为判断烟气是否溢出的有效判据，当室外补风气流速度大于临界风速时，烟气发生溢出形成外部烟气，反之则不会形成。

Spreading characteristics of external smoke for buildings with internal corridor

Abstract: The influence of external smoke under different fire characteristic parameters was analyzed in order to analyze the spreading characteristics of the external smoke. One building with internal corridor was analyzed through numerical simulation. Results show that the external smoke will re-enter the adjacent floor and result in the significant increase of temperature in fire initial stage. Then the temperature will decrease and unchange due to the decrease of external smoke volume. When the fire scale is small, the fire location has no significant effect on the external smoke. When the fire scale is higher, the influence of fire location has the dangerous area. If the fire location is in the dangerous area, the impact degree of external smoke will significantly increase and the degree of that will decrease if the fire location is out of the dangerous area. The dangerous area is 4.5m to 6.5m away from the vent. The critical velocity of wind can be the criterion that determines whether smoke spills out. If the wind velocity is higher than the critical velocity, the smoke will spill out; if not, the smoke will not spill out.

范文分析：

研究目的：In order to analyze the spreading characteristics of the external smoke

研究方法：The influence of external smoke under different fire characteristic parameters was analyzed. One building with internal corridor was analyzed through numerical simulation.

研究结论：Results show that the external smoke will re-enter the adjacent floor and result in the significant increase of temperature in fire initial stage. Then the temperature will decrease and unchange due to the decrease of external smoke volume. When the fire scale is small, the fire location has no significant effect on the external smoke. When the fire scale is higher, the influence of fire location has the dangerous area. If the fire location is in the dangerous area, the impact degree of external smoke will significantly increase and the degree of that will decrease if the fire location is out of the dangerous area. The dangerous area is 4.5m to 6.5m away from the vent. The critical velocity of wind can be the criterion that determines whether smoke spills out. If the wind velocity is higher than the critical velocity, the smoke will spill out; if not, the smoke will not spill out.

本摘要和范文2类似，只说明了研究目的、研究方法和研究结论，而且研究目的和研究方法出现在一句话中。本摘要节选自《浙江大学学报（工学版）》第53卷第10期（2019年10月），论文作者：张晓涛，陆愈实，陆凯华。

范文4：

最小剪力系数及其调整方法对超高层建筑地震响应的影响

摘要：超高层结构地震剪力响应由振型分解反应谱法得到的结果经常不能满足规定的最小剪力系数要求。为此，文章简述剪力系数的概念和调整方法，以具有不同剪力系数的两个模型对比分析结构弹性、弹塑性地震响应差异，探讨剪力系数对超高层结构地震响应的影响。以通过强度和刚度调整使最小剪力系数满足规范要求的两个模型，分析不同调整方法引起的结构响应的合理性。结果表明：满足最小剪力系数的结构的弹性基底剪力大、层间位移角较小，结构的弹塑性位移响应也较小，受力状态优于不满足最小剪力系数的结构，安全性得到了提高。结构弹性倾覆力矩需求和弹塑性基底剪力按刚度调整大于按强度调整；结构弹塑性最大顶点位移和层间位移角响应相差不大，但出现刚度大、层间位移角也大的与抗震理论相悖的情况；在满足抗震要求的情况下，构件的受力状态则是按强度调整更优，构件截面更加经济合理。

Influence of Minimum Shear Force Coefficient and Its Adjustment Methods on the Seismic Response of Super High-rise Buildings

Abstract：Seismic shear force response of super high-rise structures, calculated by the mode superposition response spectrum method, cannot meet the requirements of the minimum shear force coefficient in code. Therefore, the concept and adjustment method of minimum shear force coefficient were sketched in this paper. Two models with different shear force coefficients were then adopted to analyze their differences in structural elasticity and elastoplastic seismic response, and discuss the influence of shear coefficient on the seismic response of super-tall buildings. Finally, the minimum shear force coefficient of the two models was adjusted to satisfy code requirements through strength and stiffness adjustment methods; rationality of the structural response induced by different methods was then analyzed. Results demonstrated that, compared with the structure without a specified minimum shear force coefficient, the elastic base shear force of the structure with a specified minimum shear force coefficient was larger, and story drift ratio and elastoplastic displacement responses were smaller, i.e., the safety of the structures was enhanced. The elastic overturning moment demand and elastoplastic base shear values of the two models, adjusted by the stiffness adjustment method, were greater than those adjusted by the strength adjustment method. Using the two methods, the maximum vertex displacement and story drift ratio were similar, although there was an abnormal situation wherein the story drift ratio changed with stiffness. In terms of meeting seismic requirements, stress condition of the components was superior and the component cross-section more reasonable and economical by using the strength adjustment method.

范文分析：

研究背景：Seismic shear force response of super high-rise structures, calculated by the mode superposition response spectrum method, cannot meet the requirements of the minimum shear force coefficient in code.

研究目的和方法：Therefore, the concept and adjustment method of minimum shear force coefficient were sketched in this paper. Two models with different shear force coefficients were then adopted to analyze their differences in structural elasticity and elasto-plastic seismic response, and discuss the influence of shear coefficient on the seismic response of super-tall buildings. Finally, the minimum shear force coefficient of the two models was adjusted to satisfy code requirements through strength and stiffness adjustment methods; rationality of the structural response induced by different methods was then analyzed.

研究结果：Results demonstrated that, compared with the structure without a specified minimum shear force coefficient, the elastic base shear force of the structure with a specified minimum shear force coefficient was larger, and story drift ratio and elastoplastic displacement responses were smaller, i.e., the safety of the structures was enhanced. The elastic overturning moment demand and elastoplastic base shear values of the two models, adjusted by the stiffness adjustment method, were greater than those adjusted by the strength adjustment method. Using the two methods, the maximum vertex displacement and story drift ratio were similar, although there was an abnormal situation wherein the story drift ratio changed with stiffness. In terms of meeting seismic requirements, stress condition of the components was superior and the component cross-section more reasonable and economical by using the strength adjustment method.

　　本篇摘要有研究背景、研究目的、研究方法和研究结果部分。研究背景非常简短，研究目的和方法混在一起了，研究结果比较长。本摘要节选自《地震工程学报》第41卷第5期（2019年10月），论文作者：王贵珍，谭潜，魏俊彪，王丽萍。

练习题

请把下面的句子翻译成英语。

　　1. 随着工业4.0和中国制造2025计划的实施，数字化设计与制造智能化成为了机械发展的主要方向。

　　2. 计算机辅助公差设计（CAT）技术作为CAD/CAM集成化的重要内容，对于提高设计效率、实现智能制造具有重要意义。

　　3. 通过实验分析发现，所提出的图像识别分析方法在一定程度上可以识别出损毁建筑物，但仍需要进一步研究，以提高其识别精度。

　　4. 物联网的开放性和多样性导致其容易受到各种恶意的攻击。为了有效维护特定区域物联网网络空间安全，针对物联网中典型的窄带物联网（Narrow Band Internet of Things, NB-IOT），提出了一种基于中间人的物联网终端探测和识别技术。

　　5. 为更有效地纠正建筑工人危险认知偏差，预防不安全行为发生，以认知心理学为基础，先采用心理测量方式获取工人的危险认知结果，再以对比方式反映偏差程度，最后对偏差成因进行探索分析。

　　6. 本设计利用组态软件WinCC进行监控，通过网关节点连接到温湿度监测节点和设备测控节点，实现风机、阀门等设备的控制与环境参数监测，构建了完整的三防控制系统。

　　7. 实验表明，用LORA无线网络可以取代现有的基于PLC控制的传统三防控制系统，简化了系统的建设安装，降低了维护成本。

　　8. 通过实验验证，该技术探测和识别概率较高，具有一定的应用前景。

　　9. 本文对碳纳米管吸波材料的复介电常数和复磁导率随碳纳米管含量的变化进行探究。

　　10. 随着我国科学技术不断发展，计算机技术的发展也衍生出了人工智能技术，特别是在大数据支持下，最大程度上发挥了人工智能技术的优势，在计算机网络技术中发挥着巨大作用。

参考文献

[1] Aull L. First-year University Writing: A Corpus-based Study with Implications for Pedagogy [M]. Palgrave Macmillan, 2015.

[2] Charles M. Argument or evidence? Disciplinary variation in the use of the Noun that pattern in stance construction [J]. English for Specific Purposes, 2007 (2).

[3] Chen Y H, Baker P. Lexical bundles in L1 and L2 academic writing [J]. Language Learning & Technology, 2010 (2).

[4] Granger S, Leech G. Learner English on Computer [M]. Routledge, 2014.

[5] Groom N. Pattern and meaning across genres and disciplines: An exploratory study [J]. Journal of English for Academic Purposes, 2005 (3).

[6] Harwood N."Nowhere has anyone attempted… In this article I aim to do just that": A corpus–based study of self–promotional I and we in academic writing across four disciplines [J]. Journal of Pragmatics, 2005 (8).

[7] Hyland K. Disciplinary Discourse: Social Interactions in Academic Writing [M]. London: Longman, 2000.

[8] Murray R. Writing for Academic Journals (Second edition) [M]. McGraw-Hill Education (UK), 2009.

[9] Sinclair J M, Coulthard M. Towards an Analysis of Discourse: The English Used by Teachers and Pupils [M]. Oxford Press, 1975.

[10] 曹晟, 王昌栋, 邢建春, 等. 基于LORA物联网的人防工程三防控制系统 [J]. 计算机与控制系统, 2019, 38 (10): 118~121.

[11] 杜雪玲. 基于语料库的科技论文摘要体裁分析 [D]. 大连: 大连理工大学, 2008.

[12] 高广运, 耿建龙, 毕俊伟, 等. 地铁环境振动对建筑场地影响实测分析 [J]. 工程地质学报, 2019, 27 (5): 1116~1121.

[13] 耿照新. 科技英语写作教程 [M]. 北京: 北京交通大学出版社, 2014.

[14] 郭汉丁, 张印贤, 陈思敏. 既有建筑节能改造市场主体各阶段社会责任共担演化机理 [J]. 土木工程与管理学报, 2019, 36 (5): 25~32.

[15] 郭悦, 王红军. 基于中间人的物联网终端探测和识别 [J]. 电讯技术, 2019, 59 (10): 1197~1202.

[16] 韩豫, 尹贞贞, 刘嘉伦, 等. 建筑工人的危险认知偏差特性及成因 [J]. 土木工程与管理学报, 2019, 36 (5): 56~67.

[17] 胡庚申. 英语论文写作与发表 [M]. 北京: 高等教育出版社, 2000.

[18] 黄国文, 葛达西, 张美芳. 英语学术论文写作 [M]. 重庆: 重庆大学出版社, 2004.

[19] 李哲, 王文龙, 张梦, 等. 碳纳米管材料低频电磁参数及吸波产热特性 [J]. 化工学报, 2019, 70 (S1): 28~34.

[20] 刘定操, 尚展垒. 利用遥感图像对震损建筑结构变形检测的识别研究 [J]. 地震工程学报, 2019, 41 (5): 1380~1384.

[21] 于强, 鹿院卫, 张晓盼, 等. 纳米粒子对熔盐复合蓄热材料热物性的影响 [J]. 化工学报, 2019, 70 (S1): 217~225.

[22] 张文娟, 张海涛. 大数据时代人工智能在计算机网络技术中的运用 [J]. 网络信息工程, 2019 (9): 73~74.

[23] 王贵珍, 谭潜, 魏俊彪, 等. 最小剪力系数及其调整方法对超高层建筑地震响应的影响 [J]. 地震工程学报, 2019, 41 (5): 1161~1169.

[24] 杨娜, 崔雅萍, 任海棠. 大学英语师生课堂语码转换实证调查 [J]. 西安外国语大学学报, 2012, 20 (3).

[25] 张晓涛, 陆愈实, 陆凯华. 内廊式建筑火灾外部烟气蔓延规律 [J]. 浙江大学学报 (工学版), 2019, 53 (10): 1986~1993.

[26] 张秀国. 英语专业毕业论文写作教程 [M]. 北京: 清华大学出版社, 2007.

[27] 赵方舟, 罗大兵, 陈达, 等. 基于SolidWorks的计算机辅助公差优化设计研究 [J]. 机械设计与制造, 2019 (10): 15~19.

[28] 左志宇, 谭洁, 毛罕平, 等. 基于物联网的微型植物工厂智能监控系统设计 [J]. 农机化研究, 2019 (11): 74~79.

4 文献综述

4.1 概述

4.1.1 文献综述的简介

除本书以上篇幅中所提及的内容,我们还必须对文献综述有清晰地认识。本章将主要从文献综述的概念类型、如何写文献综述以及相应的范文赏析三大方面展开介绍。

文献综述(Review)可以是学术论文的一个组成部分,也可以是学术论文的一类,即作者在搜集大量相关文献的基础上,通过综合分析与评价,整理归纳而成的专题性学术论文。它们的写作技巧有相通之处,但是作为一篇论文,又有其独到的地方。从字面意思理解,"综"即综合,就是要求对文献资料综合分析、概括整理,使材料更加精练明确,逻辑分明;"述"即评述,要求对整理归纳后的材料进行比较专门的、系统的、全面的、深入的论述。总而言之,文献综述即对文献资料的综合和评述,是作者本人对某一方面课题的历史背景、前人工作、争论焦点、研究现状和发展前景等内容进行评论的科学性论文。好的文献综述,应有较完整的文献资料,有眼光独到的分析评论,除此之外,还能准确地反映文章主题,有合理的发展预测。

综述的类型可分为背景式综述、历史性综述、理论式综述、方法性综述及整合式综述五大类。背景式综述是最普遍的一种,其主要介绍某一问题的研究意义、背景情况,读者可以从此类综述中了解到该研究问题与前期研究的相关性,认识到前期研究中存在的错误和不足;历史性综述是一种用于追溯某一思想或理论形成和发展过程的介绍性综述,阅读此类综述,读者可以对某一学科的全貌形成一个基本的认识和了解;理论式综述主要是介绍和比较同一现象的不同理论,分析不同理论的优缺点,并比较各理论的解释力,当研究者需要拓展某一理论或者整合两种及以上理论时,常会用到此类综述;方法性综述是对相关课题中研究方法的使用进行分析和评价,指出不同的研究方法可能会导致不同的研究结果;整合式综述是研究者整理与某一研究问题相关的文献资料,为读者呈现出该研究的现状。

文献综述的内容和形式多种多样,篇幅不一。那么好的文献综述应当具备哪些特点呢?第一,文献综述要具备综合性,以时间发展为纵轴,反映当前课题的发展;以区域发展为横轴,进行横向比较。只有这样才能全方位地进行综合分析,归纳整理,从而得出更精练、更明确、更有逻辑和层次的材料,把握本课题的发展规律和预测发展趋势。第二,文献综述要具有评述性,是指对某一方面的问题进行专门地、全面地、系统地、深入地论述,对某一研究问题有自己的见解和评论。第三,文献综述还要具备先进性,即作者要搜集最新的文献资料,把最新的研究成果和发展趋势呈现给读者。

4.1.2 文献综述的研究意义和目的

我们阅读文献，目的在于超越而不是单纯地学习，进而文献综述应在归纳总结大量文献的基础上，论证别人没有提出过的观点或对他们的结论提出具有建设性的意见。故文献综述的研究意义主要在于需要解决以下问题：

（1）文献综述应当给读者展现出一幅纵观所选的全景图，包括该领域的发展历程、趋势以及是否达到过某种共识或存在过某种争议；

（2）应当点明当下研究与以往研究的关系，即该领域是否存在明显的研究空白或尚待修正的错误；

（3）应该为作者的研究问题和假设做铺垫、做论证，这主要指前人的研究理论、研究设计及方法是否存在某种不足和局限性；

（4）应当指出以往研究所选取的方面以避免重复，点明该领域的哪些方面已经做得非常充分、成熟，哪些方面还未得到足够的重视和探索。

撰写文献综述通常出于某种需要或目的，比如撰写资助申请书、撰写学位论文、描述和解释现有的知识以指导专业实践、寻找有效的研究和开发方法、寻找能够帮助解释现有文献的专家以及寻找未发表的信息源、寻找资金来源和正在开展的工作、满足个人的好奇心等。

4.1.3 文献综述的写作

文献综述的简介、研究目的及意义属于理论篇的内容，接下来我们将重点介绍实践篇——文献综述的写作，这也是我们本章的重点所在。其主要步骤大致可分为：

（1）选择主题。首先是发现自己的研究兴趣，这主要来源于人们日常生活中的一些争论焦点、矛盾问题和对某些现象的好奇心；其次是将研究兴趣具体化和聚焦，我们通过选择和简化，将目光集中于一个特定且可分析的研究对象身上；最后我们就可以根据所选的研究兴趣确定自己的研究主题。

（2）文献搜索。在确定研究课题之后，我们就可以进行文献的搜索工作。第一步为发现需要审阅的文献，在数据库中漫无目的地寻找自己想要的文献无异于大海捞针，而主题决定了研究的范围及方向，以主题陈述为线索去寻找对自己写作有用的文献。第二步则为文献的查询，不同的信息需要从不同的参考资料中获取，快速阅读文献以了解主要内容，在查询文献时要做好记录和整理，这样既可以提升自己的写作效率，又可以训练自己的表达能力和阅读水平。

（3）文献整理。文献整理为整个写作过程中承上启下的过渡部分，大致分为三步：首先是文献的阅读和筛选，对搜集来的文献先进行粗读，即重点阅读文献的摘要和结论部分，判断其是否可用。对于选用的文献要进行精读，重点阅读其研究目的、方法、主要结论和观点等。其次是文献的分类和归纳，对于搜集并选用的文献，应根据其主要研究方法、结论和观点等内容的不同，进行合理的划分和总结，使文献内容逐渐系统化并形成初步的构思。最后则为分析与写作提纲，通过对资料的综合分析和初步构思，确定综述的论证方法，安排层次结构，整理写作大纲。

（4）综合撰写。作为一类完整的学术论文，文献综述一般包括前言、主体、总结和参考文献四部分。在撰写文献综述时可先按这四部分拟写提纲,再根据此提纲进行撰写,最后对综述进行总体的修改,得到一篇完整的综述。

文献综述的写作不仅步骤严谨明确,还需要花费大量的时间和精力,同时需要注意很多写作中的问题,这是由科研本身的特性所决定的,我们将在本章的以下篇幅中重点介绍。

4.2 文献评估与利用

全面的文献调研是科学研究的基础,在大量的文献调研的基础上,才能通过去伪存真、去粗取精,推陈出新地开展具有自己特色的科学研究。因此,文献的评估和利用是保证文献综述质量的关键工作。"评估"即检验一篇文献质量的好坏,是否可用,"利用"则为文献的查找、整理和分类。其与文献综述写作步骤中的选择主题和文献搜索相对应。

4.2.1 文献评估

在确定研究主题之后,我们会通过各种渠道搜集大量的文献资料,对这些资料我们不能盲目地使用,必须进行合理的评估与筛选。检验一篇文献质量的好坏,是否可用,最基本的一点是文献主要内容是否与自己的研究主题相对应,具体来说有以下标准:

(1) 文献论述是否可信且有条理;
(2) 所涉及的文章选题是否已清晰地流出;
(3) 研究成果在多大程度上受到引用;
(4) 研究思路是否清晰可信并经过他人的认可和追随;
(5) 研究结论是如何一步一步被分析出来的;
(6) 该研究的意义及重要性如何;
(7) 该研究的发展趋势如何;
(8) 该研究的假设是否可行以及如何实现;
(9) 该研究的方法论部分是否为当前公认的研究此项成果最恰当的方法论。

那么怎样才能搜集到高质量的文献呢?从方法层面上我们要做到两点,第一要从实用的角度,去寻找那些与主题相关的,以我们能理解的语言写成的和发表在受推崇杂志上的文献;第二要从方法学质量的角度,去寻找那些最符合学者及科学家赖以收集证据的文献。

4.2.2 文献利用

在知道需要查找何种文献之后,文献的具体查找、整理和分类则成为接下来的关键工作。

4.2.2.1 文献的查找

我们从方法层面上讲述了查什么样的文献,概括的讲可以查找本领域核心期刊的文献;本领域带头人或主要课题组的文献;高引用频率的文章。此外,文献的选择应当遵循由近及远的原则,因为最新的研究常常包括之前的研究资料,可以使人更快地了解知识和认识现状。据此,我们可以用以下方法具体的去查找文献:

(1) 通过主题词、关键词检索，选择恰当的关键词和主题词才能保证检索出的文献内容丰富全面。

(2) 通过检索某个作者，查 SCI 可以知道在此领域有建树的学者，查找他近期发表的文章。

(3) 通过参考综述检索，如果有与自己的研究课题相关或贴合的综述，可以根据其相应的参考文献去寻找那些原始的研究论文。

(4) 注意文献的参考价值，文章的被引次数、刊物的影响因子虽然能反映文章的参考价值，但同时也要注意引用此文章的其他文章对其是如何评价的。

(5) 不同的信息需要从不同的参考资料中获取，应当考虑清楚自己需要哪种类型的信息，以及在哪些数据库中能够找到此类文献。

4.2.2.2 文献的整理和分类

大部分情况下大家会搜集到很多文献，而如何把这些资料整理归纳成自己的"数据库"显得尤为重要。对于大量的未读文献，可以用以下三点来解决：

(1) 关注文章前言的最后一部分，这部分一般会给出作者进行此项工作的原因、依据和方法。

(2) 关注文章的图表，了解所采用的表征方法。

(3) 关注文章的结论，是否实现了既定目标以及是否需要改进和完善。

然后尽可能的用 50 字左右来概括文章的目的、表征手段和结论，提炼出文献的主要内容。对于文献中的重要观点或自己阅读时的感悟、启示和想法应随时做好记录，以便在之后的写作过程中选用。

在进行文献整理归纳时做好笔记，而后再根据笔记将文献分类。分类的依据有很多，比如中文/英文、研究论文/综述、表征手段或结论的不同以及文献的重要程度等。

4.3 文献回顾与写作

本书 4.2 节介绍了文献的评估和利用，主要讲述查找什么样的文献，怎样查找文献以及如何对文献进行整理和分类等问题。本节内容则主要关于文献的回顾与写作，即如何对文献进行阅读、分析、归纳、整理和综合撰写，接下来我们将介绍其具体的做法。

4.3.1 文献回顾

4.3.1.1 阅读和筛选

阅读分为精读和粗读，粗读主要关注文献的摘要和结论，了解文献的主旨，看文献内容是否具有相关性、可靠性、代表性，以判断其是否可用。对于选用的文献，要进行精读，了解其目的、方法、结论和主要观点等，做好文献摘录卡。具体的阅读方法如下：

(1) 明确文献的阅读目的。文献有不同的读法，但最重要的是我们能否概括出一篇文献到底论述了什么，否则就是白读。在上节文献整理分类时就已经概括了各文献，故下面介绍几种不同目的的文献读法。

1) 回顾重要信息的读法，即阅读完文献后，回忆一下文章的重要信息是什么，如果不知道就从文献的摘要和结论部分查找，最好在讨论部分确认一下。如果你读完一篇文献下来发现需要记的东西很多，那往往是没有抓住重点。

2）扩充知识面的读法，这主要读文献的介绍部分，看他人提出的问题以及目前的进展，每天读几篇，一个月内就可以对该领域的某个方向形成大概的了解。

3）写文章的读法，重点阅读文献的讨论部分，对其中比较精练优美的英文句型做好摘抄工作并有意识的记一下，潜移默化中我们便会逐渐提高自己的写作水平和效率。

（2）根据文献的类型确定阅读顺序。如果接触到一个陌生的研究领域，我们看文献的顺序应为中文综述—中文博士论文—英文综述—英文期刊文献。

中文综述是我们快速了解自己的研究领域的入口，可以帮助我们更好地认识课题。除此之外，通过中文综述可以了解该领域的基本名词、基本参量、表征方法等，这将对我们阅读英文文献时大有帮助。同时，中文综述包含大量的英文参考文献，这就为我们后续查阅文献打下了良好的基础。

中文博士论文的第一章前言或者是绪论部分所包含的信息量普遍大于一篇综述的。因为它对该领域的背景以及相关理论的介绍更加详细，还会经常提到国内外在该领域拔尖的科研小组的相关研究方向。通过阅读可以更清楚地理清研究主题的脉络。

英文综述，尤其是发表在高 IF 期刊上或是那种 invited paper，往往都是该领域的大家所写。对于此类文献要精读，分析文章的层次和构架，要特别关注作者对各方向优缺点的评价以及对缺点的改进和展望。除此之外，我们还可以从英文综述中学到很多地道的英文表达和专业名词，对我们写作能力的提升有很大的帮助。

（3）阅读文献中各个部分。

注重摘要：摘要是文献的窗口，多数文章在浏览完题目和摘要之后即可掌握大部分内容。阅读文献时一定要注意不可走极端，既不能只看摘要，更不能过分的追求原文。

通读全文：在读第一遍时，尽量不要查字典以防思维变得混乱，最大限度地从文章中获取有用信息，将不明白的单词先放在原文中去理解，然后做好生字的标记，待通读全文后再去查找其意思。

归纳总结：论文的句子一般都比较长，容易遗忘，可以用一个词组作为标题标注每一句或者每一段。

抓住主题，确立句子架构：读英文文献的窍门在于准确理解大量的关系连词并抓住每一段落的主旨，这样可以帮助我们在阅读文献时起到事半功倍的效果。

增加阅读量：如果刚刚接触某一领域，对一些基本名词或概念不是很了解，读文献会非常吃力，但是随着阅读量的增加，你会发现事情慢慢变得很简单。

4.3.1.2 分类与归纳

对于所搜集到的文献，应根据主要方法、结果、结论及观点的不同，进行分类和归纳，使文献内容系统化并产生初步判断，形成初步构思。

4.3.1.3 分析与写提纲

在对文献进行综合分析后，要确定研究主题的论证方法，安排层次结构，整理写作提纲。对于一篇完整的论文而言，提纲的重点在于前言部分的内容和正文的各级标题，并把相应的文献标记在标题下。最后，还要检查资料是否充分可靠，观点与资料是否一致，各部分的构架是否合理等，如果存在问题，及时进行改进和补充。

4.3.2 文献写作

本部分讲述文献综述类论文写作模式。文献综述注重向读者介绍与研究主题相关的详细资料、动态发展、未来展望以及对上述各方面的评述。因此文献综述类的论文形式相对多样化，但从总体上看一般的综述都包括以下四部分：前言、主体、总结和参考文献。在撰写综述时，可先按此四部分拟写大纲，再据此进行综合撰写。以下我们将具体介绍文献综述的这四部分内容及写作中需要注意的问题。

4.3.2.1 文献综述的组成

A 前言：点题

说明此综述写作的目的和意义，介绍相关概念、定义及综述的范围，说明文献资料来源，概述所选主题的历史背景、发展历程、研究现状、争论焦点、实践意义和应用价值，使读者对文章有一初步印象，篇幅在300字左右。

B 主体

主体是综述的核心部分，主要包括：概念、定义、理论基础；研究意义和目的；主题的起源、历史背景、发展现状、争论焦点；研究方法、技术、结论及相关问题的分析和比较；研究主题的发展趋势等。主体以研究目的、表征手段及结论为主要线索，具体地向读者介绍此文章的研究课题。那么主体应该如何去写呢？

（1）明确写法。

1）纵式：（历史发展纵观）适用于动态型综述。以某一主题为中心，按照专题自身或者时间的发展顺序，对其历史发展、现状趋势预测作纵向综述，进而勾勒出某一专题的发展轨迹全景图。此类综述要求脉络清晰，对各阶段的发展动态（问题、成果、趋势）作简明扼要的描述，重点介绍突破性、创造性成果，略一般性材料。

2）横式：（国际国内横揽）适用于争鸣性、成就性综述。介绍国内外对本专题的研究现状及各派观点、各家之言，并对各种方法、结果、结论、观点等进行分析和比较，借此起到指导、启示、警戒的作用。

3）纵横结合式：写历史背景采用纵式，写现状采用横式，纵横结合。学术论文的综述大多采用纵横结合式写法，此写法可以全面系统地认识某一课题及发展动向，作出清晰可靠的发展预测，为新的研究工作提供方向和突破口。

（2）合理组织、层次分明。将综述的主要内容划分成若干层次，并相应地列出主要标题和小标题，在各级标题下列出拟叙述和讨论的内容及应准备的文献资料，拟写出文章的大概轮廓。

C 结语

对全文进行综合概括，得出自己的结论或给出建议，要求简明扼要、重点突出，强调作者最希望让读者接受的内容。

D 参考文献

参考文献也是文献综述的重要组成部分，它不仅给出了引用文献的依据以及表示对原文献作者的尊重，而且为后来者想要深入探讨该课题提供了文献检索来源。参考文献应该注意编排条目清楚、查找方便、内容准确。

根据提纲综合撰写文献综述的全部内容之后,就可以进行最后一步的修改工作,进而得到一篇质量较好的文献综述。

4.3.2.2 文献写作中应当注意的问题

文献综述的引言部分不仅要交代清楚本课题的研究背景、本文献综述的研究目的和意义,还要对文章的主要内容作出扼要说明,考虑到整体的行文衔接和过渡,以达到导读的目的。

(1) 概念应该如何写。在文献综述中需要介绍一个特定的概念时会有不同的定义,不要以作者为线索对概念进行综述。最好的做法是:先叙述这些定义在本质上有何区别,分别从什么角度进行的定义,被大众广泛接受的定义是什么,以这个定义的话题(空巢老人、女生节有什么变化发展等问题),我们应当从这些方面进行综述。综述时要对这些定义的核心思想和共同观点进行分析和归纳,还可以分析各种定义划分的依据,各种划分依据是否合理等。最好能在分析归纳各种定义的基础上提出自己的概念。

(2) 如何对文献进行回顾。我们在文献综述中列举某一问题,并给出一些学者的研究时,不仅要清晰地给出各种文献观点,还要在每个观点之后给出文献的出处,方便读者理解和以后的查阅。

(3) 如何解决文献综述中的各种逻辑问题。

1) 为了避免文献重复,最好先根据各种文献与自己研究课题的相关性进行归纳分类,再以行文中所涉及的问题来安排文献。

2) 合理安排文献的顺序,切忌一上来就给结论。作者应先对国内国外的学者的研究文献进行分析,结合文献给出的重要数据和事实,最后在此基础上给出自己结论性的观点。

3) 以关注的线索安排文献,在每个问题下给出国内外的研究成果并进行分析和比较。

4) 行文衔接和过渡。在文献综述中的每一句话尤其是具有结论性的语言之前,都要给出依据,使行文更加顺畅严谨。

5) 文章的归纳和总结。在归纳文章内容或总结各派观点时应注意连接词的使用,使文章层次分明、逻辑清晰、结构严谨。

6) 当作者按照时间顺序来行文时,常常会对时间段进行划分。此时需要说明这样划分的理论依据,依据来源于什么参考文献。切忌"蜻蜓点水"式的文献回顾。

(4) 文献综述的题目要和研究问题相结合。比如"抗日战争时期的民族政策研究"这一题目,作者的研究中心应是"民族政策",而作者的文献综述却是"抗日战争",这种文献综述的题目和研究问题当然是不匹配的。

(5) 切忌文献罗列。文献回顾并不是将文献中的研究发现和结论逐一列举出来,这样没有任何的意义。相反,作者应围绕自己的研究问题,对相关文献进行有机的归纳总结,还可以在此基础上给出自己的评价和意见,让读者可以看出所依据文献演进的内在逻辑,遵循"文献树"的写作原则。

(6) 关于引用文献的问题。

1) 当需要注明所引用文献的作者时,一般做法是再次引用时只需列出文献的第一作者。而对于两个作者,每次引用文献时都要给出两个作者的姓名。

2) 文章中出现的重要数据都要给出。

3) 当引用文献时，时间、页码、作者、期刊应有所标注。一般的做法是作者+时间。

4) 在引用《古代汉语词典》《美国联邦贸易委员会法案》等文献时，需要给出具体信息，比如作者、出版社和出版年月日等。对于引用的法案、法律、法规等，则需要给出具体的出版时期和颁布的部门机构等细节信息。

（7）切忌文献综述缺乏权威性。我们在学术研究中必须坚持学术标准，不能掺杂有学术标准之外的任何因素。因此应选取权威人士的权威著作，这可以丰富自己文献综述的理论依据，提升文章可信度。

（8）切忌直接引用另一篇文献综述中的内容。如果在搜集文献时发现了一篇与自己研究问题很贴合的文章，切忌直接引用其内容。正确的做法是查找原始文献，了解原始文献中有哪些学者进行了相关研究，其后再对这些文献进行归纳和分类，再进行综述。

（9）文献综述中提出假设要给出理由。我们不能简单地罗列出几个有关的研究实例后就给出自己的研究假设，这会降低自己研究的价值和意义。正确的做法是在回顾文献的基础上，发现以往的研究结论在不同的条件下可能会出现不同的结果，或是发现以往研究中所存在的问题和不足，随后再提出自己的假设。

4.4 范文阅读与分析

4.4.1 文献综述范文导读

一篇文献综述的文章如何写？应该从几方面综述？综述文章都包含哪些部分？下文用一篇文献综述文章来进行分析导读。

Example：

A Literature Review of Security Attack in Mobile Ad-hoc Networks[①]

【Abstract】Security is a major concern for protected communication between mobile nodes in a hostile environment. In hostile environments adversaries can bunch active and passive attacks against intercept able routing in embed in routing message and data packets. In this paper, we focus on fundamental security attacks in Mobile adhoc networks. MANET has no clear line of defense, so, it is accessible to both legitimate network users and malicious attackers. In the presence of malicious nodes, one of the main challenges in MANET is to design the robust security solution that can protect MANET from various routing attacks. However, these solution are not suitable for MANET resource constraints, i.e., limited bandwidth and battery power, because they introduce heavy traffic load to exchange and verifying keys. MANET can operate in isolation or in coordination with a wired infrastructure, often through a gateway node participating in both networks for traffic relay. This flexibility, along with their self-organizing capabilities, are

---

① Goyal P, et al. A literature review of security attack in mobile ad-hoc networks [J]. International Journal of Computer Applications, 2010, 9: 12.

some of MANETs biggest strengths, as well as their biggest security weaknesses. In this paper different routing attacks, such as active (flooding, black hole, spoofing, wormhole) and passive (eavesdropping, traffic monitoring, traffic analysis) are described.

【Keywords】 Security Attack; Literature Review

1 Introduction

In [1, 3, 4, 6] Mobile Ad Hoc Networks (MANETs) has become one of the most prevalent areas of research in the recent years because of the challenges it pose to the related protocols.

MANET is the new emerging technology which enables users to communicate without any physical infrastructure regardless of their geographical location, that's why it is sometimes referred to as an "infrastructure less" network. The proliferation of cheaper, small and more powerful devices make MANET a fastest growing network. An ad hoc network is self organizing and adaptive. Device in mobile ad hoc network should be able to detect the presence of other devices and per form necessary set up to facilitate communication and sharing of data and service. Ad hoc networking allows the devices to maintain connections to the network as well as easily adding and removing devices to and ROM the network. The set of applications for MANETs is diverse, ranging from large-scale, mobile, highly dynamic networks, to small, static networks that are constrained by power sources. Besides the legacy applications that move from traditional infrastructure environment into the ad hoc context, a great deal of new services can and will be generated for the new environment. It includes:

(1) Military Battlefield;
(2) Sensor Networks;
(3) Medical Service;
(4) Personal Area Network.

Security solutions are important issues for MANET, especially for those selecting sensitive applications, have to meet the following design goals while addressing the above challenges.

MANET is more vulnerable than wired network due to mobile nodes, threats from compromised nodes inside the network, limited physical security, dynamic topology, scalability and lack of centralized management. Because of these vulnerabilities, MANET are more prone to malicious attacks. The primary focus of this work is to provide a survey on various types of attacks that affect the MANET behavior due to any reason.

2 Related Work

A MANET is a most promising and rapidly growing technology which is based on a self-organized and rapidly deployed network.

Due to its great features, MANET attracts different real world application areas where the networks topology changes very quickly. However, in [4, 7] many researchers are trying to remove main weaknesses of MANET such as limited bandwidth, battery power, computational power, and security. Although a lot of work under progress in this subject particularly routing attacks and its existing countermeasures. The existing security solutions of wired networks cannot be applied directly to MANET, which makes a MANET much more vulnerable to security attacks. In this paper, we have discussed current routing attacks in MANET.

Some solutions that rely on cryptography and key management seem promising, but they are too expensive for resource constrained in MANET. They still not perfect in terms of trade offs between effectiveness and efficiency. Some solutions in [4, 7, 12] work well in the presence of one malicious node, they might not be applicable in the presence of multiple colluding attackers. In addition, some may require special hard ware such as a GPS or a modification to the existing protocol.

The malicious node(s) can attacks in MANET using different ways, such as sending fake messages several times, fake routing, information, and advertising fake links to disrupt routing operations. In the following subsection, current routing attacks and its countermeasures against MANET protocols are discussed in detail.

3 Manet Vulnerabilities

A vulnerability is a weakness in security system. A particular system may be vulnerable to unauthorized data manipulation because the system does not verify a user's identity before allowing data access. MANET is more vulnerable than wired network. Some of the vulnerabilities are as follows:

3.1 Lack of Centralized Management

MANET doesn't have a centralized monitor server. The absence of management makes the detection of attacks difficult because it is not east to monitor the traffic in a highly dynamic and large scale ad-hoc network. Lack of centralized management will impede trust management for nodes.

3.2 Resource Availability

Resource availability is a major issue in MANET. Providing secure communication in such changing environment as well as protection against specific threats and attacks, leads to development of various security schemes and architectures.

Collaborative ad hoc environments also allow implementation of self organized security mechanism.

3.3 Scalability

Due to mobility of nodes, scale of ad hoc network changing all the time. So scalability is a major issue concerning security.

Security mechanism should be capable of handling a large network as well as small ones.

3.4 Cooperativeness

Routing algorithm for MANETs usually assume that nodes are cooperative and non-malicious. As a result a malicious attacker can easily become an important routing agent and disrupt network operation by disobeying the protocol specifications.

3.5 Dynamic Topology

Dynamic topology and changeable nodes membership may disturb the trust relationship among nodes. The trust may also be disturbed if some nodes are detected as compromised. This dynamic behavior could be better protected with distributed and adaptive security mechanisms.

3.6 Limited Power Supply

The nodes in mobile ad hoc network need to consider restricted power supply which will cause several problems. A node in mobile ad hoc network may behave in a selfish manner when it is find that there is only limited power supply.

4 Security Goals

Security involves a set of investments that are adequately funded.

In MANET, all networking functions such as routing and packet forwarding, are performed by nodes themselves in a self organizing manner. For these reasons, securing a mobile ad hoc network is very challenging. The goals to evaluate if mobile ad hoc network is secure or not are as follows:

4.1 Availability

Availability means the assets are accessible to authorized parties at appropriate times. Availability applies both to data and to services. It ensures the survivability of network service despite denial of service attack.

4.2 Confidentiality

Confidentiality ensures that computer-related assets are accessed only by authorized parties. That is only those who should have access to something will actually get that access. To maintain confidentiality of some confidential information, we need to keep them secret from all entities that do not have privilege to access them. Confidentiality is sometimes called secrecy or privacy.[5]

4.3 Integrity

Integrity means that assets can be modified only by authorized parties or only in authorized way. Modification includes writing, changing status, deleting and creating. Integrity assures that a message being transferred is never corrupted.

4.4 Authentication

Authentication enables a node to ensure the identity of peer node it is communicating with. Authentication is essentially assurance that participants in communication are authenticated and not impersonators. Authenticity is ensured because only the legitimate sender can produce a message that will decrypt properly with the shared key.

4.5 Nonrepudiation

Nonrepudiation ensures that sender and receiver of a message cannot disavow that they have ever sent or received such a message. This is helpful when we need to discriminate if a node with some undesired function is compromised or not.

4.6 Anonymity

Anonymity means all information that can be used to identify owner or current user of node should default be kept private and not be distributed by node itself or the system software.

5 Security Attacks

Securing wireless ad hoc networks is a highly challenging issue.

Understanding possible form of attacks is always the first step towards developing good security solutions. Security of communication in MANET is important for secure transmission of information.[4] Absence of any central coordination mechanism and shared wireless medium makes MANET more vulnerable to digital/cyber attacks than wired network, there are a number of attacks that affect MANET. These attacks can be classified into two types:

5.1 Passive Attacks

Passive attacks are the attack that does not disrupt proper operation of network. Attackers snoop data exchanged in network without altering it. Requirement of confidentiality can be violated if an attacker is also able to interpret data gathered through snooping. Detection of these attack is difficult since the operation of network itself does not get affected.

5.2 Active Attacks

Active attacks are the attacks that are performed by the malicious nodes that bear some energy cost in order to perform the attacks. Active attacks involve some modification of data stream or creation of false stream. Active attacks can be internal or external.

5.2.1 External attacks are carried out by nodes that do not belong to the network.

5.2.2 Internal attacks are from compromised nodes that are part of the network.

Since the attacker is already part of the network, internal attacks are more severe and hard to detect than external attacks. Active attacks, whether carried out by an external advisory or an internal compromised node involves actions such as impersonation (masquerading or spoofing), modification, fabrication and replication.

6 Active Attacks

6.1 Black Hole Attack

In this attack, an attacker advertises a zero metric for all destinations causing all nodes around it to route packets towards it[9]. A malicious node sends fake routing information, claiming that it has an optimum route and causes other good nodes to route data packets through the malicious one. A malicious node drops all packet that it receive instead of normally forwarding those packets. An attacker listen the requests in a flooding based protocol.

6.2 Wormhole Attack

In a wormhole attack, an attacker receives packets at one point in the network, "tunnels" them to another point in the network, and then replays them into the network from that point. Routing can be disrupted when routing control message are tunneled. This tunnel between two colluding attacks is known as a wormhole. In DSR, AODV this attack could prevent discovery of any routes and may create a wormhole even for packet not address to itself because of broadcasting. Wormholes are hard to detect because the path that is used to pass on information is usually not part of the actual network. Wormholes are dangerous because they can do damage without even knowing the network.

6.3 Byzantine Attack

A compromised with set of intermediate, or intermediate nodes that working alone within network carry out attacks such as creating routing loops, forwarding packets through non-optimal paths or selectively dropping packets which results in disruption of degradation of routing services within the network.

6.4 Rushing Attack

Two colluded attackers use the tunnel procedure to form a wormhole. If a fast transmission path (e.g. a dedicated channel shared by attackers) exists between the two ends of the wormhole, the tunneled packets can propagate faster than those through a normal multi-hop route. The rushing attack can act as an effective denial-of-service attack against all currently proposed on-demand MANET routing protocols, including protocols that were designed to be secure, such as ARAN and Ariadne[14].

6.5 Replay Attack

An attacker that performs a replay attack are retransmitted the valid data repeatedly to inject the network routing traffic that has been captured previously. This attack usually targets the freshness of routes, but can also be used to undermine poorly designed security solutions[8].

6.6 Location Disclosure Attack

An attacker discover the Location of a node or structure of entire networks and disclose the privacy requirement of network through the use of traffic analysis techniques[10], or with simpler probing and monitoring approaches[14]. Adversaries try to figure out the identities of communication parties and analyze traffic to learn the network traffic pattern and track changes in the traffic pattern. The leakage of such information is devastating in security.

6.7 Flooding

Malicious nodes may also inject false packets into the network, or create ghost packets which loop around due to false routing information, effectively using up the bandwidth and processing resources along the way. This has especially serious effects on adhoc networks, since the nodes of these usually possess only limited resources in terms of battery and computational power.

Traffic may also be a monetary factor, depending on the services provided, so any flooding which blows up the traffic statistics of the network or a certain node can lead to considerable damage cost.

6.8 Sinkhole

In a sinkhole attack, a compromised node tries to attract the data to itself from all neighboring nodes. So, practically, the node eavesdrops on all the data that is being communicated between its neighboring nodes. Sink hole attacks can also be implemented on ad hoc networks such as AODV by using flaws such as maximizing the sequence number or minimizing the hop count, so that the path presented through the malicious node appears to be the best available route for the nodes to communicate.

6.9 Spoofing Attack

In spoofing attack, the attacker assumes the identity of another node in the network; hence it receives the messages that are meant for that node. Usually, this type of attack is launched in order to gain access to the network so that further attacks can be launched, which could seriously cripple the network. This type of attack can be launched by any malicious node that has enough information of the network to forge a false ID of one, its member nodes and utilizing that ID and a lucrative incentive, the node can misguide other nodes to establish routes towards itself rather than towards the original node.

6.10 RERR Generation

Malicious nodes can prevent communications between any two nodes by sending RERR messages to some node along the path.

The RERR messages when flooded into the network, may cause the breakdown of multiple paths between various nodes of the network, hence causing a no. of link failures.

6.11 Jamming

In jamming, attacker initially keep monitoring wireless medium in order to determine frequency at which destination node is receiving signal from sender. It then transmit signal on that frequency so that error free receptor is hindered.

6.12 Replay Attack

The attacker collects data as well as routing packets and replays them at a later moment in time. This can result in a falsely detected network topology or help to impersonate a different node identity. It can be used to gain access to data which was demanded by replayed packet.

6.13 Sybil Attack

The Sybil attack especially aims at distributed system environments. The attacker tries to act as several different identities/nodes rather than one. This allows him to forge the result of a voting used for threshold security methods. Since ad hoc networks depend on the communication between nodes, many systems apply redundant algorithms to ensure that the data gets from source to destination. A consequence of this is that attackers have a harder time to destroy the integrity of information

6.14 Sinkhole Attack

The attacking node tries to offer a very attractive link e. g. to a gateway. Therefore, a lot of traffic by passes this node. Besides simple traffic analysis other attacks like selective forwarding or denial of service can be combined with the sinkhole attack.

6.15 Desynchronization Attack

In this attack, the adversary repeatedly forges messages to one or both end points which request transmission of missed frames.

Hence these messages are again transmitted and if the adversary maintains a proper timing, it can prevent the end points from exchanging any useful information. This will cause a considerable drainage of energy of legitimate nodes in network in an end-less synchronization-recovery protocol.

6.16 Overwhelm Attack

In this attack, an attacker might overwhelm network nodes, causing network to forward large volumes of traffic to a base station. This attack consumes network bandwidth and drains node energy.

6.17 Blackmail

A black mail attack is relevant against routing protocols that uses mechanisms for identification of malicious nodes and propagate messages that try to blacklist the offender.

6.18 Denial of Service Attack

Denial of service attacks are aimed at complete disruption of routing information and therefore the whole operation of ad hoc network.

6.19 Gray-hole Attack

This attack is also known as routing misbehavior attack which leads to dropping of messages. Gray hole attack has two phases.

In the first phase the node advertise itself as having a valid route to destination, while in second phase, nodes drops intercepted packets with a certain probability.

6.20 Selfish Nodes

In this a node is not serving as a relay to other nodes which are participating in the network. This malicious node which is not participating in network operations, use the network for its advantage to save its own resources such as power.

6.21 Man-in-the-middle Attack

An attacker sites between the sender and receiver and sniffs any information being sent between two nodes. In some cases, attacker may impersonate the sender to communicate with receiver or impersonate the receiver to reply to the sender.

6.22 Fabrication

The notation "fabrication" is used when referring to attacks performed by generating false routing messages. Such kind of attacks can be difficult to identify as they come as valid routing constructs, especially in the case of fabricated routing error messages, which claim that a neighbor can no longer be contacted [5].

6.23 Impersonation

Impersonation attacks are launched by using other nodes identity, such as IP or MAC address. Impersonation attacks are sometimes are the first step for most attacks, and are used to launch further, more sophisticated attacks.

7 Passive Attacks

7.1 Traffic Monitoring

It can be developed to identify the communication parties and functionality which could provide information to launch further attacks. It is not specific to MANET, other wireless network such as cellular, satellite and WLAN also suffer from these potential vulnerabilities.

7.2 Eavesdropping

The term eavesdrops implies overhearing without expending any extra effort. In this intercepting and reading and conversation of message by unintended receiver take place.

Mobile host in mobile adhoc network shares a wireless medium.

Majorities of wireless communication use RF spectrum and broadcast by nature. Message transmitted can be eavesdropped and fake message can be injected into network.

7.3 Traffic Analysis

Traffic analysis is a passive attack used to gain information on which nodes communicate with each other and how much data is processed.

7.4 Syn Flooding

This attack is denial of service attack. An attacker may repeatedly make new connection request until the resources required by each connection are exhausted or reach a maximum limit. It produces severe resource constraints for legitimate nodes.

8 Conclusion

In this paper, we have analyzed the security threats an ad hoc network faces and presented the security objective that need to be achieved. On one hand, the security-sensitive applications of an ad hoc networks require high degree of security. On the other hand , ad hoc network are inherently vulnerable to security attacks. Therefore, there is a need to make them more secure and robust to adapt to the demanding requirements of these networks.

The flexibility, ease and speed with which these networks can be set up imply, they will gain wider application. This leaves ad hoc networks wide open for research to meet these demanding application. The research on MANET security is still in its early stage. The existing proposals are typically attack-oriented in that they first identify several security threats and then enhance the existing protocol or propose a new protocol to thwart such threats. Because the solutions are designed explicitly with certain attack models in mind, they work well in the presence of designated attacks but may collapse under unanticipated attacks.

Therefore, a more ambitious goal for ad hoc network security is to develop a multi-fence security solution that is embedded into possibly every component in the network, resulting in depth protection that offer multiple line of defense against many both known and unknown security threats.

References

[1] Chiang C C. Routing in Clustered Multihop, Mobile Wireless Networks with Fading Channel, Proc. /EEe SICON97, Apr. 1997, 197~211.

[2] Clausen Th, et al. Optimized Link State Routing Protocol, IETF Internet draft, draft-ietf-manet-olsr-11. txt, July 2003.

[3] Kannhavong B, Nakayama H, Nemoto Y, et al. A survey of routing attacks in mobile ad hoc networks. Security in wireless mobile ad hoc and sensor networks, October 2007, 85~91.

[4] Karakehayov Z. Using REWARD to Detect Team Black-Hole Attacks in Wireless Sensor Networks, Wksp. Real-World Wireless Sensor Networks, June 20~21, 2005.

[5] Desilva S, Boppana R V, Mitigating Malicious Control Packet Floods in Ad Hoc Networks, Proc. IEEE Wireless Commun. and Networking Conf, New Orleans, LA, 2005.

[6] Lee S, Han B, Shin M. Robust Routing in Wireless Ad Hoc Networks, 2002 Int'l. Conf. Parallel Processing Wksps. , Vancouver, Canada, Aug. 18~21, 2002.

[7] Kurosawa S. et al. Detecting Blackhole Attack on AODV-Based Mobile Ad Hoc Networks by Dymamic Learning Method, Proc. Int'l. J. Network Sec., 2006.

[8] Johnson D, Maltz D. Dynamic Source Routing in Ad Hoc Wireless Networks, Mobile Computing, T, Imielinski and H. Korth, Ed, 153~81, Kluwer, 1996.

[9] Jyoti Raju Garcia-Luna-Aceves J J. A comparison of On-Demand and Table-Driven Routing for Ad Hoc Wireless etworks, in Proceeding of IEEE ICC, June 2000.

[10] Hu Y C, Perrig A, Johnson D. Wormhole Attacks in Wireless Networks, IEEE JSAC, vol. 24, no. 2, Feb. 2006.

[11] Al-Shurman M, Yoo S M, Park S. Black Hole Attack in Mobile Ad Hoc Networks, ACM Southeast Regional Conf, 2004.

[12] Zapata M G, Asokan N. Securing Ad Hoc Routing Protocols, Proc. 2002 ACM Wksp. Wireless Sec., Sept. 2002, 1~10.

[13] Sanzgiri K, et al. A Secure Routing Protocol for Ad Hoc Networks, Proc. 2002 IEEE Int'l. Conf. Network Protocols, Nov. 2002.

[14] Perkins C, Royer E. Ad Hoc On-Demand Distance Vector Routing, 2nd IEEE Wksp. Mobile Comp. Sys. and Apps., 1999.

[15] Yi P, et al. A New Routing Attack in Mobile Ad Hoc Networks, Int'l. J. Info. Tech., vol. 11, no. 2, 2005.

（1）从总体上讲，《A Literature Review of Security Attack in Mobile Ad-hoc Networks》是一篇结构完整的文献综述，既有综述又有评论。文章开头交代了写作背景、目的及意义；主体部分则采用横式写法（国内国际横揽），介绍了对于移动节点之间的安全是保护通信的研究；最后部分对开发嵌入到网络中可能的每个组件中的多栅栏安全解决方案进行分析和评价。

（2）在行文结构中作者有明确的思路，文章的整体篇章结构是以研究专题为线索来进行划分，即按照概念界定、表现特征、产生原因、解决对策来行文。

（3）关于文章的结构划分如能给出理由依据更好。此外，在写文献综述的过程中可以适当使用图表，这可以使结果更加直观易懂。

（4）在引用国内外文献时，可以在对各派观点、各家之言描述的基础上进行分析和比较，总结出各种见解、方法、观点的优劣利弊，最好能够给出自己的观点。

（5）在最后的总结评价部分，如能给出自己的建设性意见或者关于本专题的发展预测会使文章更加全面丰富。

4.4.2 文章中文献综述部分的写法导读

文章中的文献综述部分也称文献回顾，其写法与文献综述文章有所不同，这部分如何撰写？应该包含哪些内容？下文用两个范例进行导读。

Example 1：(Deformation Failure Characteristics of Coal Body and Mining Induced Stress Evolution Law❶): With the deterioration of occurrence condition of coal resources, mechanical environment for coal mining, organization structure of coal and rock mass, its mechanical behavior

---

❶ Wen Zhijie, et al. Deformation failure characteristics of coal body and mining induced stress evolution law [J]. The Scientific World Journal, 2014.

and failure characteristics become complicated, so the characteristics of engineering response change. Meanwhile, the large space mining model complicates time-space relationship and kinetics features of mining stress field rock burst and coal and gas outburst, which causes serious damage and a number of casualties. The main reasons for these problems are lack of knowledge of the space-time evolution law of mining stress field with different mining conditions, roadway excavation and maintenance, and the advance of heading face at the wrong time and space.

The domestic and foreign scholars usually use classical mechanics, (e. g, elastic-plastic mechanics) to study the damage development process of coal body and the mining stress field that loads on the coal body during the process under the condition of mining activity. In this analytical method, ideal coal body is under the condition of static stress. But coal body in the stoep is under dynamic loading induced by the bending and breaking of the overlaying strata, which weakens the mechanical properties of coal body and continuously transfer the mining stress inside the coal body. So the quantitative study on the space-time evolution mechanism and its law of mining stress field with the deterioration development of mechanical properties of coal body is necessary. It is helpful to the roadway excavation, maintenance, and the advance of the heading face, which has very important practical significance.

作者在综述时思路明确，明确主干线索，先经过举例，提出概念，而后总结研究方法、现实意义来行文。

前半部分通过对各种风险因素的举例介绍，引出了对砂矿演化规律变化的概念，另一部分进而讨论了国内外学者的研究方法，以及实验条件下煤体材料的物理力学性质，总结出一系列经验方法，有非常大的推广前景。

Example 2: With increased mining depths and intensities, rock burst, a kind of dynamic disaster, seriously threatens coal mining safety. Consequently, domestic and foreign experts and field engineers have performed significant research on controlling rock bursts, and many prevention and control technologies have been proposed. Among them, stress-relief-blasting technology is the most efficient technology for preventing rock bursts. After stress-relief-blasting, the properties of coal are changed, and the performance effectiveness of the stored energy in coal is reduced. This principle is very important in preventing and controlling rock bursts resulting from stored energy mechanisms and can be used to reduce the occurrences of rock bursts.

Among the multiple factors influencing blasting effect, charging structure is one of the key factors influencing energy transfer and rock breakage. Charging structures include the concentration and position of explosives in the blast hole and the coupling condition between the explosives and the blast-hole wall. There are two kinds of charging structures: coupling and non-coupling charge structures. In coupling charge structures, explosives fill the entire radial space of the blast hole, whereas in non-coupling charge structures, some radial gap exists between the explosives and blast hole. Anon-coupling charge structure increases the production expansion of coal fractures and reduces stress concentrations in coal-rock mass, thus maximizing the use of explosive energy.

Many studies have been performed on stress-relief blasting and non-coupling charges. Yang et al used an explosion load-test system involving digital laser dynamic caustics to analyze the dynamic-fracture effects of backscratching with different charging structures. By analyzing everlasting stress wave characteristics of rock fragmentation, Xu et al gave formulas for the accurate calculation of fissure diameter under a stress wave. Using a numerical simulation method, Wei et al analyzed the stress distribution and variation rules of rock before and after stress-relief blasting and verified the effects of stress relief in mining engineering combined with electromagnetic radiation monitoring technology. Gao et al studied the attenuation law of the mechanical properties of rock by conducting a different test. Through laboratory and field blasting tests, Guo studied the parameters of blasting crack spread and blast-hole spacing. Guo studied the different loads exerted on a hole by solid-liquid coupling and non-coupling blasting through laboratory experiments.

These studies mainly focus on the common rule of rock blasting. However, because each rock burst disaster may vary, targeted control techniques have a more practical significance. To guide the practical control of rock bursts, this study analyzes the stress-relief blasting zone of coal from the characteristics of rock bursts. In addition, a laboratory experiment was conducted to determine the characteristics of non-coupling blasting, and the effects of stress-relief blasting on specimens were analyzed through acoustic emissions (AE) and stress monitoring. Finally, deep non-coupling blasting was adopted for roadways with high stress concentrations.

作者在综述时明确主干线索，先行介绍措施，引出措施施效机理，进而说明控制因素，最后得出结论的方法来行文。之前先通过介绍煤炭开采环境病害防治措施，来引出具体作用机制，介绍病害发生机理与具体操作原理，最后作出结论，具体问题具体分析，强调每一种方法都有其具体应用特点的全文线索来行文。

练习题

1. 什么是文献综述？你的理解？
2. 文献综述的写作一般分为哪几个步骤？
3. 如何理解"接触到一个陌生的研究领域，我们看文献的顺序应为中文综述—中文博士论文—英文综述—英文期刊文献"？

参考文献

[1] Goyal P, et al. A literature review of security attack in mobile ad-hoc networks [J]. International Journal of Computer Applications, 2010, 9: 12.
[2] Wen Zhijie, et al. Deformation failure characteristics of coal body and mining induced stress evolution law [J]. The Scientific World Journal, 2014.

5 论文主体写作

本部分是论文的主体，即表达作者的研究成果，主要阐述自己的观点及其论据，包括引言、正文（材料与方法、结果与讨论）、结论。这部分要以充分有力的材料阐述观点，要准确把握文章内容的层次、大小段落间的内在联系。篇幅较长的论文常用推论式（即由此论点到彼论点逐层展开、步步深入的写法）和分论式（即把从属于基本论点的几个分论点并列起来，一个个分别加以论述）两者结合的方法。

主体结构包括以下几个方面：

（1）引言（Introduction）：这个章节要回答的问题是"为什么"（Why）。论文开头首先描述这篇论文进行的研究是想要解决什么问题，并解释该问题的重要性。

（2）材料与方法（Materials and Methodology）：这部分要回答的问题是"怎么研究"（How）。文中应提供足够的实验细节，包含材料的来源、设备型号和制造商、数量、实验持续时间和季节等，让其他人可遵循文中的细节进行实验复制。所有的研究都应具有可重复性，因此这一部分也就特别重要。

（3）结果（Findings or Results）：这个章节应回答"什么"（What），即通过研究你的新发现是什么？只谈结果，把评论和说明放到讨论章节。使用适当的表格和图片，但不要重复数据，如相同数据在表格和图片上重复或图片里数据重复在内文中提及。

（4）讨论（Discussion）：此章节要回答的最重要的问题是"所以呢?"（So what）。解释研究结果意味着什么，为什么重要。与先前研究结果做对比，并解释任何矛盾。如果某些结果没达到统计显著关系，解释观察到的差异的可能，列出研究的局限性，并提出今后可进行的工作。最后，总结结论。

（5）结论（Conclusion）：这部分是论文的归结收束部分，要写论证的结果，做到首尾一贯，同时要写对课题研究的展望，提及进一步探讨的问题或可能解决的途径等。

5.1 引言（Introduction）

与中文论文"简短"的"概述"（或"前言"）不一样，英文的引言内容通常较长。好的论文在这部分很见功底，文献的阅读量、信息综合能力，可以给读者很多的信息量，因此写好它容不得半点马虎。

引言部分的内容通常用来为作者创造一个研究空间。先介绍目前的研究现状，然后指出存在的不足或尚没有解决的问题，最后再介绍"存在的问题"是"如何"被作者的研究所解决。因此，由以下三部分来组成。

（1）提出研究现状和此研究的重要性。先通过陈述表明所要研究问题的重要性——当然这部分内容不是必须，并介绍此领域的研究现状，具体可参考文献综述引用。

研究问题要与自己的研究内容高度相关，时态一般可用一般现在时，并通过很确定的语气和具体的形容词来强调研究的重要性。

The flow of foams is seen in many process, and its use in major industries means that an understanding of foam rheology is of paramount importance.

（2）强调有必要解决存在的问题。指出该研究目前存在的问题，可以通过提问的方式或者通过某种方式扩展此领域已有知识和结论。

这一部分非常重要，只有指出存在的问题或尚待解决的问题，才能突显出自己的研究价值。在这一部分的写作时，一般通过转折词来表示过渡，并在指出问题时使用负面的词汇。

…; however, the relationship between emergence and soil temperature has not been investigated previously…

In contrast to the extensive literature describing…, little attention has been paid to…

（3）介绍作者自己的研究内容。介绍作者的研究目的和大致的研究内容。也可以在此部分表明自己的研究假设前提，宣布自己的主要研究发现、它们的价值，如有需要，也可以介绍论文的框架结构——方便读者了解复杂的文章结构。

这部分的内容主要用来介绍作者如何"填补"提出的"问题"或"不足"。

This paper presents the results of an extensive experimental program…

关于此处的时态，比如是 presents 还是 presented？即是一般过去时还是一般现在时？一般而言：

1）如果主语是 paper, article, thesis, report 等指代文章本身的抽象词汇，使用一般现在时；

2）如果主语是实验调查本身，比如 study, experiment, investigation 等具体词汇，使用一般过去时或一般现在时都可以。

万能法则：此处的时态如果拿不准，一直使用一般现在时。

除了以上三部分，还需要注意以下事项：

（1）引言结构如"倒金字塔"，即从一个"宽泛的研究领域"到一个"本文要做的特定的东西"。结构不能颠倒，而且"倒金字塔"的"基底"不能无限宽广，即：不能从漫无边际的地方说起。

（2）在介绍别人和自己做了什么前期工作时，要根据这一小部分的功能，有针对性、概括性地综述，不能"堆积材料"。所谓"堆积材料"就是说引用每一篇文献时用了大量篇幅进行描述，没有概括出和本文有关的要点，而是把该文献的摘要抄了一遍。还有的人写作时，为了引文献而引文献。

（3）避免使用时髦语，避免滥用套话。时髦语如"黄金催化剂的研究是当今催化研究皇冠上的钻石，是北极星"。看到这样的话，审稿人就会举起大斧砍下来。另外，避免滥用套话的意思就是说引言部分不能到处都是"什么什么课题引起广泛兴趣，在什么什么中有广泛应用"，如果每一段都这么说，就无趣了。

（4）指出课题组前文和本文的联系和区别。回答既然前面已经发表过几篇文章了，为什么本文还是值得发表？既然已经发过一篇快报，现在为什么要发长文章？和前文相比，有哪些新方法、新内容、新理论？是否达到更好的应用效果？如果在引言部分写清楚这些东西，审稿人的思维就会根据这些东西来判断是否可信，是否值得发表。

(5) 在引言部分的末尾建议用简练的话把本文的重要结果"预览"一下。但注意不能使用大篇幅预览，不能把正文里面的信息详细透露，一览无余。

5.1.1 引言的常见写法

论文的引言非常重要，一个好的开端能吸引读者的注意力。因此，写好论文的引言是很关键的一步。一篇论文如何开头？不同的作者和不同的文章处理开端的方式不同，但常见的写法有以下几种：

(1) 背景法。说明研究的背景资料等。

例：在过去的40年中，有关微波传输线特性阻抗计算研究工作已做了很多。到目前已有很多种计算方法可供使用，这些方法可以分为以下三类……

In the past 4 decades, considerable work has been done on the calculation of characteristic impedance for the microwave transmission line. A number of calculation methods are available, classified into the following three kinds：…❶

以背景法开头的写作有：

- In recent years, there has been a wide concern in…
- It is noteworthy that…has made progress…
- In most studies of…, …has been emphasized with attention being given to…
- Several researchers have theoretically investigated…

(2) 主题句法。提出一个观点或者论点作为论文要阐明和论述的主题。

例：本文阐述了利用偏微分方程方法，求得复杂目的的射频场解。

The solution of RF field problems for objects of realistic complexity by partial differential equation methods are presented in this paper.

以主题句开头的写法有：

- The present research deals mainly with…
- This paper focuses on…
- In this paper, …is discussed (studied, investigated, described, etc)
- The emphasis of this paper is to survey…

(3) 问题法。用提出问题的方法来引出文章的内容，以吸引读者的注意力。

例：What types of land use affect mode choice? In particular what is the role of density and commercial land use?

- What can be found from…?
- Does it hold water that…?
- What is function/role of…?

5.1.2 引言中常见的问题

引言中常见的问题如下：

(1) 篇幅不当。引言在论文中起定向引导作用，因此不宜太长。有些作者在写作过程

---

❶《电子科学期刊》，Vol. 7, No. 3, 1990.

中不知道如何合理安排引言包含的内容，如何将它们有条有理地给读者描述清楚，便将与论文研究相关的资料全部写入，致使引言篇幅过长；有些作者为了突出自己的工作，将引言写的过分简单，致使读者不知道本文研究的内容有什么意义。为了合理控制引言的篇幅（字数），通常我们在写作过程中将引言分为四到五个层次，即课题研究的背景、意义、研究现状、当前研究过程中存在的问题，本文欲研究的内容等，然后根据各部分承担的任务与发挥的功能，运用简洁、明了的语言将论文要表述的内容表述清楚，一般应控制在200字左右，占正文的 1/10~1/8。

（2）和摘要内容雷同。从内容上看，摘要是论文的缩影，是全文的高度概括，其要素主要是目的、方法、结果和结论，其中内容和结果是重点，其主要功能是提供给读者论文的主要内容，同时便于计算机检索，以更好地为科技工作者提供服务。而引言则是说明论文研究的必要性，重点写课题研究的背景意义、分析研究现状、归纳当前亟需解决的问题，目的是引导读者理解全文。从结构上看，摘要是一篇完整的短文，可以独立于论文之外而存在；而引言则是论文的一个重要组成部分，缺少引言直接进行正文研究内容的叙述，则会让读者感到课题研究的展开没有依据，研究的必要性自然会遭到质疑。因此，作者在科技论文引言写作过程中应注意认真分析二者的各自特点和功能，不能将其简单地与摘要重复，也不能写成摘要的注释。

（3）和正文内容混淆。很多作者在引言写作过程中加入一些论文内容中比较重要的插图列表、公式的推导证明过程、化学方程式以及分子结构式等。这些内容是为了直观、形象地说明论文正文中的某一问题，所以应该将其安排在正文中，而引言部分则是作者提出问题的部分，因此不能将其列入引言中。

（4）与结论重复或二者没有建立关系。还有一些作者不明白引言和结论各自承担的功能，将引言写成结论，还有些作者不能兼顾论文结构的整体性，对于引言中提出的问题，在结论中不给予解答，或者前后不能照应，引言讲一方面的问题，结论却讲另一方面的现象。例如，引言中讲到了前人的工作，结论却没有说明作者的研究工作哪些与前人相同，在前人的基础上又通过什么方法对此进行了改进，改进的结果怎样，还有哪些方面需要进一步解决等。通常情况下，引言是提出正文要研究的问题，给读者以引导；而结论则是整篇论文的总结和概括，对提出的问题给予解答，因此，两者不可等同，也不可不建立联系。

（5）背景论述繁简不当。论文研究的意义和必要性即是论文研究的背景，通常与某项研究成果在某一领域的应用有密切联系，一般情况下需要作者广泛阅读文献来获得必要信息，因此面对当今世界海量的信息，如何筛选必要的信息就显得十分重要了。一些作者喜欢将其描述得非常详细，好像文章的综述，一些作者认为研究背景和意义都是目前已经有的东西，无须着重论述，为此将背景论述的非常简单甚至不提供背景材料，这样均失去引言存在的意义。

5.1.3 引言实例

文题：NaCl 胁迫对黄瓜种子萌发及幼苗生长的影响❶

---

❶ 杨秀玲，郁继华，李雅佳，等. NaCl 胁迫对黄瓜种子萌发及幼苗生长的影响［J］. 甘肃农业大学学报，2004，39（1）：6.

黄瓜在我国蔬菜设施栽培中占有重要地位。由于设施栽培采用特殊的覆盖结构，改变了黄瓜生长的内部生态环境及自然状态下的水热平衡，尤其是大大改变了土壤的理化性质，致使土壤次生盐化的程度越来越高，给黄瓜周年生产造成巨大损失。目前，关于作物盐胁迫效应方面的研究多集中于粮食作物、造林植物以及蔬菜作物中的番茄、黄瓜、西瓜、佛手瓜等，而对黄瓜的这方面研究报道较少。本研究采用人工控制盐分浓度的方法，研究了 NaCl 胁迫对黄瓜种子的萌发特性以及黄瓜幼苗的生长、根系活力、细胞保护酶活性及膜脂过氧化的影响，为黄瓜的栽培、耐抗盐性的筛选和育种工作提供理论和技术依据。

Cucumber plays an important role in vegetable facility cultivation in China. The special mulch structure adopted in the facility cultivation has changed the internal ecological environment of cucumber growth and the hydro thermal balance in the natural state, especially the physical and chemical properties of the soil, resulting in the increasing degree of secondary salinization of the soil and causing huge losses to the annual production of cucumber（目前研究对象存在的问题，本研究的理由和背景）. At present, studies on the effects of crop salt stress mainly focus on food crops, afforestation plants and vegetable crops such as tomato, spinach, watermelon, citron, etc., while studies on this aspect of cucumber are less reported（介绍研究对象基本特征）. Adopting the method of artificial control of salt concentration, this paper focuses on the effect of the NaCl stress on cucumber seed germination, the growth of cucumber seedlings and roots and cell protective enzyme and membrane lipid peroxidation, which offers theoretical and technical base concerning the cultivation, resistance to salt screening and breeding work of cucumber（本研究成果和意义）.

这篇引言言简意赅，重点突出，以简短篇幅介绍了论文的写作目的和相关领域研究工作的概况，点出了本文研究的依据和内容，引出主题。

5.2　正文（Body）

正文是科技论文的核心组成部分，占全文的主要篇幅。如果说引言是提出问题，正文则是分析问题和解决问题，回答"怎么研究"（How）这个问题。这部分是作者研究成果的学术性和创造性的集中体现，它决定着论文写作的成败和学术、技术水平的高低。正文应充分阐明科技论文的观点、原理、方法及具体达到预期目标的整个过程，并且突出一个"新"字，以反映科技论文具有的首创性。根据需要，论文可以分层深入，逐层剖析，按层设分层标题。

一般地说，正文宜分为若干段落，各段落都要有自己的小主题。要写好正文必须考虑好能表达各段落中心思想的主题句，还要注意各段落中内容的一致性（Unity）和文章的连贯性（Coherence）。段落与段落之间的"承"和"转"技巧的应用，都是写作中应该注意的。

正文的目的是介绍科学或技术成果，即人们在认识某项科技领域的客观规律方面所取得的进展。这些成果都是在认识过程中,人们将丰富的感性材料加以去粗取精、去伪存真、由此及彼、由表及里的改造制作,经过实验或者调查研究后形成的观点。到了写论文的时候,人们不过是利用论文来充分地反映这些已经形成了的观点。因此,论文就自然地需要以基本观点为中心,依靠逻辑关系把整个体系串起来,同时运用反映事物本质的、内在联系的材料来说明观点。

正文的论述方式可以有两种形式:一种是将科学研究的全过程作为一个整体,对有关各方面作综合性的论述;另一种是将科学研究的全过程按研究内容的实际情况划分为几个阶段,再对各个阶段的成果依次进行论述。由于研究对象、研究方法和研究成果的不同,以及学科的不同,对正文的写作和编排不能作出统一的规定,但一般的正文部分都应包括研究的对象、方法、结果和讨论这几个部分。

试验与观察、数据处理与分析、实验研究结果的得出是正文的主要部分,应该给予有重点的详细论述。要尊重事实,在资料的取舍上不应掺入主观成分,或妄加猜测,也不应忽视偶发性现象和数据。

论文不必要讲求辞藻华丽,但要求思路清晰、合乎逻辑、用语简明准确、明快流畅。内容务求客观、科学、完备,应尽量利用事实和数据说理。凡是用简要语言能够讲述清楚的内容,应用文字陈述;用文字不容易说明白或说起来比较繁琐的,可用图或表来说明。图或表要具有自明性,即图表本身给出的信息就能够表达清楚要说明的问题。避免用图和表反映相同的数据。图和表要精心选择和设计,删去可有可无的或重复表达同一内容的图和表。引用的资料,尤其是引用他人的成果应注明出处。

切忌用教科书式的方法撰写论文,对已有的知识避免重复论证和描述,尽量采用标注参考文献的方法;对用到的某些数学辅助手段,应防止过分注意细节的数学推演,必要时可采用附录的形式供读者选阅。

正文撰写中涉及量和单位、插图、表格、数学式、化学式、数字用法、语言文字和标点符号、参考文献等,都应符合有关国家标准的要求。

5.2.1 正文内容的顺序排列

英语科技论文中的正文一般有两种写法,一种是大体上按研究工作进程的时间顺序,依次叙述。在研究工作的实践中,原来是分不同层次进行的,经过多次循环,深化了对研究内容的认识。所以,在论文正文的写作中也要按照认识问题的先后,一个问题一个问题来写。

安排问题的序列,要有其认识上的逻辑性。文章开始时,先对整个工作过程的层次应略有交代,然后,把一个问题作为一个层次内容,层层有实验结果,有小结,导出下一层次工作的引子。最终有综合,有分析,有总的观点和结论性的结果。另一种方法是按逻辑顺序排列。它是把研究工作全过程中多次由实践到理论的循环融合起来,提炼出典型的材料和观点,按认识由感性到理性的规律,逻辑地排列成章节。不是按试验工作的原有时间顺序,而是按照认识过程由低级向高级阶段的演变。一般首先介绍试验用的材料、试验设备、试验经过和试验结果。然后根据在试验过程中的步骤和实验结果,将数据和观察到的现象整理出来,加以综合,分别从实践上升到概念和判断。最后,再进行必要的讨论,归结出结论，完成推理的阶段。

按逻辑排列的正文的写作方法有两种，一种是划分法，一种是分析法，分别叙述如下。

（1）划分法。划分法是把事物划分为几个组成部分，分别予以处理。如一篇文章讲的是保护中国传统建筑的问题，可以根据内容将其分为4点，如：

Ⅰ architecture protection in the east.

Ⅱ architecture protection in the west.

Ⅲ architecture protection in the south.

Ⅳ architecture protection in the north.

然后再根据每一点内容下细分论述，如：

Ⅰ architecture protection in the east.

1. architecture protection in the past.

2. architecture protection in the present.

3. architecture protection in the future.

这种划分方便、清晰、条理，有层次，合乎逻辑，因而是科技论文正文写作的常用方法。

（2）分析法。分析法是由作者提出需要解决的问题，需要采取的措施或需要解释的现象。然后对其进行分析，直接找到解决这些问题的方法，或说明所采取的针对性措施的好处与不足，或解释清楚造成某些现象的原因；也可对几种可供选择的方法进行解释和对比，并证明为什么某种选择是最可接受的。

分析法常常用以下的方式表示：

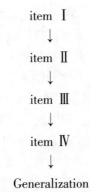

分析法实际上是归纳写作手法的体现。但是由于用其写作时对写作人员的文字功底要求较高，通常情况下若不是进行多项情况的比较，一般不宜采用。

5.2.2 主题句的写作方法（Topic Sentence）

不论长短，科技论文的正文都是由几个段落组成的。要安排好正文，分段是否得体是很重要的。

一篇论文的正文可由描写段开始，紧接着是推演段、归纳段、辩论段。多数科技论文采用推演段，作者先把一段的观点写个主题句，接着描写细节，提出支持主题句或者具体体现主题句的论证。英语论文非常强调主题句，而汉语论文并不如此。所以在撰写英语科技论文时，一定要有"主题句"意识。

主题句位置可以不固定，可以出现在段落的开端或者末尾。但写好主题句，要遵循三条原则。

（1）要有说明段落中心思想的内容。

（2）要有能体现中心思想的关键词。如在讲述"建筑物的整体平移技术新方法"的文章中，关键词是"平移技术"（monolithic moving technology）和"新方法"（new way）是关键词，因此，英文中的主题句可写为：A new monolithic moving technology has been developed concerning the building movement.

（3）尽量使用简洁明了的句子。为了直接又清晰地标明中心思想，主题句宜简不宜繁，因此不主张使用较复杂的句式和结构。

如：（×）An evaluation system has been created using the ideas which is presented in his paper.

（√）An evaluation system has been created from the ideas presented in his paper.

5.2.3 拓展句的写作方法

在写好一个段落的主题句后，就应按照主题句中的控制部分来扩展该段落的内容，以完成整个段落的写作。也就是说，要根据主题句中控制部分的内容对其进行扩展。具体地说，要对控制部分的内容要求进行讨论、说明、支撑或者证明等。那么，在扩展句中所要使用的是事实、理由、原因、说明、图表、比较、对比、定义等方式来阐述主题句中的思想内容。所以，一段写作质量较高的文字必须遵循以下三个原则。

（1）连贯性。连贯性就是要在一段文字中将所有的句子清楚地、有逻辑地并按先后顺序联系在一起，共同说明这一段的主题。在表达连贯性很强的段落中，每一句话都应自然地从前一句中繁衍而出，对该段的中心思想进行扩展。理想地说，这样的一段文字应具有一种流畅感，读者很容易找到自然的逻辑。所以，在这样的一段文字中，可采用时间顺序法、空间描述法、比较对比法以及按重要性来排列一段文字中各句的顺序等写作手法。

为了把连贯性表达清楚，可以使用的写作手法有很多，如：时间顺序法、空间顺序法、演绎法、定义法、比较法、对比法、归纳法、高潮顺序法和混合顺序法等。但是不管使用何种方法都必须首先考虑到主题句中控制部分的要求，然后再使用不同类型的连词、代词（用以指代前句中提到的名词，因为它们可以把一段中的句子与句子连接起来），以及使用必要的重复，如重复某些关键的词汇和词组，或使用同义词或词组来表示重复。这样做能使读者记住前面的内容，同时把各句也都联系起来了。

（2）完整性。如果一个段落没有完整地或充分地说明主题句所要求叙述的内容，那么这个段落就不是一个意义完整的段落。所谓完整性就是要将主题句中要求表示的内容用一致性和连贯性的方式完整地表达清楚。完整性首先应按照主题句所要求表达的内容来提供具体、翔实而富有说服力的事实。其次就是运用好各种写作技巧，使文章段落结构严谨。

（3）保持正文段落内部的一致性。段落内部的一致性体现在两个方面：行文中思维和逻辑的一致性。一段文章中应该只说明一个问题，或者一个问题的某一方面。每段只有一个中心思想或主题。所有次要句要为中心思想服务。主题句和次要句在意义上统一。如：

Structural Equation Modeling for Travel Behavior Research. （结构方程模型在出行行为上

的研究)

　　Structural equation modeling (SEM) is a modeling technique that can handle a large number of endogenous and exogenous variables, as well as latent (unobserved) variables specified as linear combinations (weighted averages) of the observed variables. (结构方程建模的内涵) Regression, simultaneous equations (with and without error-term correlations), path analysis, and variations of factor analysis and canonical correlation analysis are all special cases of SEM. (结构方程模型的特例)

　　The defining features of structural equation modeling. (结构方程建模的特征)
　　……
　　Model specification and identification. (结构方程建模在出行中的具体应用和鉴别)
　　……

　　从以上示例中体现了论文段落的逻辑性和一致性，即围绕一个问题展开，中心突出(结构方程建模在出行行为的应用研究)。

　　段落的一致性还体现在：文章中的格式、文体、专业词汇的用法以及文献列表等都需要符合一致性的标准。不要忽视任何文章中出现的不一致的成分。很多专家在审稿的时候第一就是看文章格式和文献，比如是否第一次定义缩略词，之后一致使用？是否具有一致字体、空行、子标题？文献列表格式是否一致？英语表达中的指代和单复数是否做到了一致？如：We have considered several examples and they showed good agreement with results obtained from a volume integral formulation 中 "we" 和 "they" 指代不清，主观色彩浓。应改为：Several examples have been considered and they showed good agreement with results obtained from a volume integral formulation.

5.3　结果和讨论 (Result and Discussion)

　　实验结果是对研究中所发现的重要现象的归纳，论文的讨论由此引发，对问题的判断推理由此导出，全文的一切结论由此得到。作者要指明结果在哪些图表公式中给出，对结果进行说明、解释，并与模型或他人结果进行比较。作者应以文字叙述的方式直接告诉读者这些数据出现何种趋势、有何意义。不能仅在图表中列出一大堆数据而让读者自己解读这些资料。

5.3.1　如何写结果和讨论

　　(1) 实验或计算结果和具体的分析判断应该逐项讨论。切忌堆积实验数据和结果。
　　(2) 在描述实验结果的同时，要与以前的研究结果(自己和他人的)进行对比。
　　(3) 对背后的科学问题讨论必不可少。作者在研究中得出的见解，虽然还没有充分的证据证明以作为结论的，也可以进行阐述。研究与以往研究一致不一致的结果可以阐述。如果不一致一定要尽可能给出原因。
　　(4) 结语或小结，进一步提出的新问题。陈述主要发现，特别要讨论结果中的差别、研究的意义、未解答的问题及今后的研究方向讨论，用一个句子表示较为理想。

5.3.2 结果和讨论应注意的问题

(1) 在文章中适当的给出一些具体的数值,并给人一些定量的概念。比如百分之多少、几倍等。不管实验还是理论,给出结果时要注意有效数字的位数,有效数字的位数不能超过实验或者计算的精度。

(2) 不要引证讨论文献知识太多(不同于学位论文),从而掩盖了本工作的贡献。讨论时,应该引用经典文献,一手文献,不要引用二次文献。

(3) 分析不合逻辑,结论不当。讨论太浮浅,文献知识不熟悉。避免写成工作总结,缺乏学术高度。

5.3.3 结果撰写常用句型

(1) 根据本人的研究结果做出推论。

These results suggest that the two basic AI approaches for providing decision support have been the macro, or top-down, modeling of human intelligence and the micro, or bottom-up, modeling of human intelligence.

(2) 作者解释研究结果或说明产生研究结果的原因。

These findings have shown that this design accommodates mostly rolling, rather than sliding, contact of the tooth surface.

(4) 作者对此次研究结果与其他研究者研究的结果作比较。

The recognition rate of our system is significantly higher than that reported for Token's system.

(5) 作者指出自己的理论模型是否与实验数据符合。

The measured temperatures along the heat pipe are all highly consistent with the predictions of the theoretical model.

(6) 从以上句型中可以看出学术论文在陈述研究成果时,通常使用一般现在时态。当评论的内容为对研究结果可能的证明时,句子的主要动词之前通常加上 may 或 can 等情态动词。例如:

The layer structural or some other mixed complex material may be the most suitable refrigerants for the Ericsson magnetic refrigerator.

5.3.4 讨论撰写常用句型

(1) 概述结果。

· These experimental results support the original hypothesis that …

· Our findings are in substantial agreement with those of…

· The experimental values are all lower (higher) than the theoretical predictions.

· The results given in Figure 3 validate (support) the second hypothesis.

(2) 表示研究的局限性。

· It should be noted that this study has examined only …

· The findings of this study are restricted to…

· This study has addressed only the question of…
· The limitations of this study are clear…
· The result of the study cannot be taken as evidence for…

5.4 结论（Conclusion）

论文的结论部分，是反映论文中通过实验、观察研究并经过理论分析后得到的学术见解。结论是整篇论文的结局，不是某一局部问题或某一分支问题的结论，也不是正文中各段的小结的简单重复。结论体现作者更深层的认识，且是从全篇论文的全部材料出发，以研究成果和讨论为前提，经过严密的推理、判断、归纳等逻辑分析过程而得到的新的学术总观念、总见解。结论是一篇论文的收束部分，是以研究成果为前提，经过严密的逻辑推理和论证所得出的最后结论。同时，结论部分应明确指出论文研究的成果或观点，对其应用前景和社会经济价值等加以预测和评价，并指出今后进一步在本研究方向进行研究工作的展望与设想。

5.4.1 结论涵盖的内容

（1）本文研究结果说明了什么问题。
（2）对前人有关的看法作了哪些修正、补充、发展、证实或否定。
（3）本文研究的不足之处或遗留未予解决的问题，以及对解决这些问题的可能的关键点和方向。

结论部分是全篇论文的归宿，起着画龙点睛的作用，每篇科技论文在正文后面都要有结论。虽然结论在论文中所占的篇幅不多，但写起来并不轻松。作者在结论部分的写作不符合标准要求，有很多不妥之处：结论似摘要；结论像引言；结论中出现文献引用；结论中出现图表；结论不精练；结论不具体等。

5.4.2 结论撰写应注意的问题

（1）不能模棱两可，含糊其词。用语应斩钉截铁，数据准确可靠，不用"大概""也许""可能是"这类词语，以免有似是而非的感觉，怀疑论文的真正价值。
（2）不能用抽象和笼统的语言。一般不单用量符号，而宜用量名称，比如，不说"T 与 p 呈正比关系"而说"××温度与××压力呈正比关系"。
（3）结论不能写成对文中各段小结的简单重复。如果得出的结果的要点在正文没有明确给出，可在结论部分以最简洁易懂的文字写出。
（4）不要轻率否定或批评别人的结论，也不必作自我评价，如用"本研究具有国际先进水平""本研究结果属国内首创""本研究结果填补了国内空白"等语句来作自我评价。成果到底属何种水平，读者自会评说，不必由论文作者把它写在结论里。

5.4.3 结论部分的常用句型

（1）Two factors to influence building removal project have been studied …
（2）Through the example of a 60-storey building, it has demonstrated that a simplified ap-

proach can be used …

（3）This study/This investigation/The result clearly demonstrates that…

（4）It is clear from the foregoing discussion that...

作者有时对自己研究结果采取慎重的态度，不那么肯定地说"结果说明了……""证明了……"等，用情态动词是常用的技巧。最常用的情态动词有 will，can，may 及其过去式 would，could，might 等，表示"可能"之意。

（5）Evaluation of decline symptoms in smaller size classes might provide additional useful information to be used in understanding the oak regeneration problem.

5.5　句子和段落写作中的语言问题

科技论文写作中常见的问题，除了行文中思维和逻辑不一致、文章中的格式、文体、文献列表等细节上的不一致，还要注意英语语言使用的规范，常见问题如下。

5.5.1　词汇使用不当

词汇是构成句子的最小单位。在词汇表达上，英语学术论文用词规范、严谨；多倾向于使用正式语体的词语，除非出于修辞效果上的考虑，一般不用俚语、俗语，力求给人以持重感，避免流于谐谑、轻佻；较多使用专业化技术词汇。为摆脱句子冗长、结构盘结之弊，学术英语重简明、畅达，重条理性、纪实性和充分的论据，它强调论理的客观性，不主张作者表露个人的感情，倡导在论证中排除感情因素，尽其"解释"而非"感召"之功能，因而用词正式，多使用中性词、非人格化的词。

（1）使用词的完整形式，避免使用省略形式：使用 do not 而非 don't；will not 而非 won't；it is / it has 而非 it's；there is 而非 there's；they have 而非 they've 等。

（2）使用正式或标准词，避免使用不正规或非标准的词或词组，诸如口语、俚语和忌讳性的词组。用 cannot tolerate it 代替 can't hack it；用 I do not have any idea…代替 I don't have a clue…；用 please go away 代替 flake off。

（3）使用具体的词和词组，避免使用模糊或不确切的术语。避免使用 bit，thing，stuff 等词。

（4）尽量选用单词的首要含义，避免利用单词不常见、不正式或文学含义的词。例如：

1）表示原因时，避免使用具有较强时间含义的"since"来代替"because"。

劣：Since solvent reorganization is a potential contributor, the selection of data is very important.

优：Because solvent reorganization is a potential contributor, the selection of data is very important.

2）用"and"或"or"时含义要明确，避免用斜线符号（/）。

劣：Hot / cold extremes will damage the samples.

优：Hot and cold extremes will damage the samples.

（5）避免使用双重否定表达法。如：

劣：This reaction is not uncommon.

优：This reaction is common；或：this reaction is rare；或：this reaction occurs about 40% of the time.

（6）注意起修饰、限制作用单词的位置。例如，下面的"only"：

Only the largest group was injected with the test compound. （表示 no other group）

The largest group was injected with only the test compound. （表示 no other compounds）

The largest group was injected with the only test compound. （表示 no other test compounds）

（7）常用词汇专业化（Professionalized terms）。学术写作的用词具有高度术语性，其专业术语多来源于拉丁语和希腊语。他们的意义比较稳定，利于精确地表达概念。

（8）使用中性词。避免使用"man"，代之以"people""humans""human beings""human species"等。例如：

劣：Man's search for beauty and truth has resulted in some of his greatest accomplishments.

优：The search for beauty and truth has resulted in some of our greatest accomplishments.

同样，根据表达需要用"workers""staff""work force""labor""crew""employees""personnel"等代替"manpower"；用"synthetic""artificial"等代替"man-made"。

避免使用"he""his""he or she""his or her"等，代之以"they""theirs""we""us""ours"等；尽量不使用第一人称代词"I"和第二人称代词"you"，若不能避免，用第三人称指示性的词如"the writer""the present author"等。如：

劣：The principal investigator should place an asterisk after his name.

优：Principal investigator should place an asterisk after their names.

（9）使用非人格化的词。在学术写作中，作者一般都极力避免在表达一种意见或观点时带有个人色彩，而尽量使用非人格化的词汇。例如：

It can be seen that…，而不使用 You can see that…

It is commonly believed that…，而不使用 I believe…

其他表达如：It is found through research that…

Studies have revealed that…

5.5.2 句式使用不当

学术论文句子结构较复杂，句型变化及扩展样式多。文章的逻辑性较强，具体体现在：动词一般现在时和被动态的出现频率较高，多使用陈述句和复杂句，一般不使用祈使句、反意问句和感叹句；很少使用夸张、拟人、借代、比喻等修辞手段。多数使用完全句，很少使用省略句。较多地使用形式主语 it 引导的句式和用 that 引导的主语从句；较多使用非谓语动词形式。另外，主从复合句、同位语、插入语的使用频率也较高。

（1）复杂句和长句。有时，论文写作中被描述的问题较复杂，需要使用复合句来清楚地表达，因此，各类复合句和长句在学术论文中出现的频率也较高。例如：

Debt is distinguishable from the above two categories, which together comprise "equity", in that the corporation must repay the debt holders' contribution within a certain time and debt holders are entitled to regular payment of fixed interest rather than to such dividends as the directors may declare.

句中 which 引起的非限制性定语从句，in that 引起的状语从句以及 such as 引起的定语从句。

（2）大量使用被动语态。由于学术英语以叙述某一过程、强调事实为主，句子的重点往往不在于"谁做"而在于"做什么"和"怎样做"，因此动作的执行者常处于无关紧要的地位，而且，作者在占重要语法成分的主语中可放入大量的信息。例如：

优：At welding temperature, the metal sheets apply the pressure, thus it squeezes out the oxide film and any impurities which are trapped between the sheets, and makes the weld.

劣：At welding temperature, a strong pressure is applied to the metal sheets, thus the oxide film and any impurities which are trapped between the sheets, are squeezed out, and the weld is made.

（3）it 句型及结构（it + be + adj./participle（分词）+ that clause…）。学术论文重客观事实，强调论理的客观性，不主张作者表述中带有个人主观色彩，在文献回顾、结果讨论等部分常常使用 it 句型及结构。例如：

优：It was shown that nitrite（亚硝酸盐）could be substituted for nitrate（硝酸盐）in the cure solution with the production of a more uniform and completely satisfactory product in a shorter period of time.

劣：It is apparent that the behavior of a fluid flowing through a pipe is affected by the viscosity of the fluid and the speed at which it is pumped.

（4）非谓语动词的大量使用。非谓语动词在学术论文写作中的大量运用是为了减少时间从句或表示时间的句子的使用，从而达到行文更紧凑的效果。例如：

优：Weapon is associated with violence in the mind of English spoken people, referring to the tools which could do harm to others from knives to missiles, while arms are just the fighting tools of soldiers.

优：The students of experimental group and control group participating in this experiment were strictly confidential.

（5）名词代替名词从句。有时，名词或名词短语可用来代替名词从句以达到确切、简洁的目的。例如：

优：The examination of the efficiency of the new design is necessary.

劣：It is necessary to examine whether the new design is efficient.

（6）同位语及插入成分。在学术论文写作中，同位语可用来进一步说明前面某一名词的内容，使本位语的含意具体化；同样地，插入成分的运用也是为了进一步补充说明句子的含意，使表达更明确、清晰。例如：

The owners of the corporation, the shareholders, do not as a general ruler have any personal liability for corporate debts, so that their financial exposure is said to be "limited" to their investment in the corporation.

练习题

1. 论文的主体包括几部分，它们各起什么作用？
2. 论文的引言撰写需要注意什么问题？试分析如下引言。

文题：高原鼢鼠的红细胞、血红蛋白及肌红蛋白的测定结果[1]

高原鼢鼠是一种高海拔地区的地下鼠，其洞道环境的最大特点是低氧，高 CO_2 浓度，无光，低温和潮湿。王祖望等人研究发现这种环境对地面鼠非常有害，而高原鼢鼠却能在这样的环境中世代繁衍，说明其对洞道低氧高 CO_2 浓度环境有特殊的适应机制。本文通过测定高原鼢鼠血液中红细胞数目，血红蛋白浓度以及心肌和骨骼肌细胞中肌红蛋白含量来探讨高原鼢鼠对地下生活的特殊适应机制。

The plateau zoko rat is a kind of subterranean rat in high altitude areas, whose tunnel environment is characterized by oxygen, high CO_2 concentration, darkness, low temperature and humidity. Professors such as Wang Zuwang insist that this kind of environment is hazardous to ground rats, while the plateau zoko rat can reproduce in such an environment for generations, indicating that it has a special adaptation mechanism to the environment with low oxygen and high CO_2 concentration in the tunnel. In this paper, the special adaptation mechanism of the plateau zoko rats is investigated by measuring the number of red blood cells, hemoglobin concentration and myoglobin content in the blood of the plateau zoko rats.

3. 论文的结果和讨论如何开展？试分析以下文本，举例说明。

In the existing building, additional underground space is required, and the pile supporting the upper structure is required to excavate the soil below. With the excavation of the earthwork, the stability of the pile will be reduced, and the lateral support of the pile body is an effective method to improve the stability. This paper uses ANSYS to establish a two-dimensional finite element model to analyze the variation of the stability of laterally supported piles under excavation conditions. The results show that the pile buckling ultimate load and the supporting internal force reach the maximum when a support is added at a ratio of 0.5 times the excavation value of the pile surrounding soil. The pile buckling ultimate load and the supporting internal force increase with the increase of the support stiffness. When the stiffness increases to a certain value, the increase of the load value and the internal force value is not obvious. With the increase of the support quantity, the ultimate buckling load value of the pile gradually increases, but the increase range gradually decreases. Therefore, in order to strengthen the building under excavation condition, the variation of the stability of laterally supported piles should be considered, such as a ratio which makes the supports most solid and stiff.

参考文献

[1] 李赋宁. 高级英语写作教程 [M]. 北京：外语教学与研究出版社，2008.

[2] 丁往道，吴冰，等. A Handbook of Writing [M]. 北京：外语教学与研究出版社，2001.

[3] 丁西亚. 英语科技论文写作——理论与实践 [M]. 西安：西安交通大学出版社，2006.

[4] Robert A Day, Barbara Gastel. 科技论文写作与发表教程 [M]. 6 版. 曾剑芬，译. 北京：电子工业出版社，2006.

[5] 傅温. 建筑英语论文写作 [M]. 北京：中国计划出版社，2007.

[6] 陈平，王凤池. 土木工程基础英语教程 [M]. 北京：北京大学出版社，2015.

[7] 谭家美，王岑屹. 交通工程专业英语 [M]. 上海：上海交通大学出版社，2014.

[8] 唐一平. 机械工程专业英语 [M]. 2 版. 北京：电子工业出版社，2012.

[9] 李国奇，贾强. 桩周土开挖状态带侧向支撑桩稳定性数值分析 [J]. 山东建筑大学学报，2019（6）：39~44.

[10] Day R A, Gastel B. How to Write and Publish a Scientific Paper (6th Edition) [M]. New York: Greenwood Press, 2006.

---

[1] 魏登邦. 高原鼢鼠的红细胞、血红蛋白及肌红蛋白的测定结果 [J]. 青海大学学报，2001，19（4）：1.

6 参考文献引注规范

6.1 概 述

人们对于参考文献的理解众说纷纭。从内容上看，参考文献从原来的"为撰写或编辑论著而引用的有关图书资料"[1]，到进一步定义为"为撰写或编辑论文和著作而引用的有关文献信息资源"[2]，所引内容除图书资料以外，还包括纸质或电子期刊、杂志、书评、影评、电视节目评论、政府文件、百科全书、硕博论文或网上资源等。从范围上看，广泛意义上的参考文献指作者为了写作而阅读过的所有资料，虽然作者的知识积累建立在这些资料之上，但是这些资料并不全部体现在文献引用上。狭义上来讲，参考文献是指作者在写某一著作或论文时，直接参考的文献资料，在文内和文后都有标注，且文内和文后的文献标注数量上要统一。国际英文期刊常用标准 APA 和 MLA 均要求文内引用和文后文献标注要统一。本章中所涉及参考文献，均指狭义上的参考文献。

作为学术传承的载体，文献是研究的基础。在学术论文写作过程中，对前人文献的梳理与引用是关键一环。通过梳理文献，研究者了解所研究问题的发展脉络，并通过分析比较既有的研究成果，发现研究中存在的问题和漏洞，总结前人的经验教训，从而对学术问题进行深入研究和探索，弥补之前研究中的不足并开拓新的领域。这也是学术论文选题的价值体现。同时，读者也可根据参考文献对该论文的研究基础给予评估。

在撰写论文的过程中，经常需要引用前人的研究观点或结论。为了尊重原创成果，端正学术态度，避免学术不端，营造一个良好的学术氛围，在对他人的观点或者结论进行引注时，需要遵循引注规范。引注包括参考文献的文内引用（In-text Citations）和文后的标注（MLA：Works Cited；APA：References）。英文论文参考文献引注规范常见于 MLA（Modern Language Association）格式 和 APA（American Psychological Association）格式。MLA（Modern Language Association）由美国现代语言协会制定，是一种常用的、严谨的引注格式，多用于英语论文和偏重人文学科的论文，如文化研究、文学批评等，现在已是第七版。而起源于 1929 年的 APA（American Psychological Association）格式，即美国心理学会出版的《美国心理协会刊物准则》，使用哈佛大学文章引用格式，更倾向于心理学、教育学、社会科学领域或自然学科的研究，是被广泛认可的研究论文的撰写格式。在国内，外语类期刊多借鉴 APA 格式。需注意的是，尽管 MLA 和 APA 是国际公认的英语论文格

---

[1] 全国文献工作标准化技术委员会第六分委员会. 文后信息与文献 参考文献著录规则：GB/T 7714—2005 [S]. 北京：国家标准局，2005.

[2] 中华人民共和国国家质量监督检验检疫总局，中国国家标准化管理委员会. 信息与文献 参考文献著录规则：GB/T 7714—2015 [S]. 北京：中国标准出版社，2015.

式，但是，因为不同学校和出版社的要求有异，论文格式也有所不同。

本章只涉及常见的参考文献引注格式（其他相关内容请参考 MLA Handbook for Writers of Research Papers 7th Edition、APA 国际英语论文格式的最新版本或各出版社、期刊、高校的具体要求）。本章包括概述、英语学术论文参考文献引注规范简介及部分工程类英语期刊引注格式举例、引用过程中的预防剽窃、范文阅读与分析以及最后的习题部分。本章超过四行的句段引用举例按照 MLA 格式，参考文献按照本书出版社的要求著录（但是所引用文章中的参考文献保留原有的格式），文内引用统一用呼应注且各页分别编序。

6.2 英语学术论文参考文献引注规范简介及部分工程类英语期刊引注格式举例

6.2.1 MLA 与 APA 文内引用常用格式对比

6.2.1.1 直接引用文献中的具体内容

MLA：不超过三行的直接引用部分写入正文并用引号（""）标出，引号后紧跟着用圆括号夹注（Parenthetical Citation）的形式简要注明所引文献的出处，即（作者姓氏+空格+页码），如是英文名，作者姓氏处只用"last name"；若是汉语名字，则只用姓氏的汉语拼音，不用汉字，页码直接用阿拉伯数字表示。句号、逗号、冒号、分号、问号等放在圆括号后，如果引用语句本身是疑问句或感叹句，问号和感叹号则要放到引号内。同样，文后要有相应的参考文献。如果所引文献作者为两位，则姓氏要用 and 连接；所引文献作者为三位，夹注格式为（作者1姓氏，作者2姓氏 and 作者3姓氏）。例如：

It is "by the boundary conditions at the outermost grid points" that the general cubic splines and the natural cubic splines are different (Bulant and Klims 138).

The author's own voice is of great importance, "you can use your sources as support for your insights, not as the backbone of your paper" (Hei and Guo 71).

若所引文献作者的姓氏已在正文同一句中出现，则不需要在括号夹注中重复，夹注中只需页码即可。例如：

Bulant and Klims stress that it is "by the boundary conditions at the outermost grid points" that the general cubic splines and the natural cubic splines are different (138).

Hei Yuqin and Guo Fenrong attach great value to the author's own voice, "you can use your sources as support for your insights, not as the backbone of your paper" (71).

所引内容若不在一页上，则夹注格式举例如下：(Bulant 138-140)

用方括号"[ ]"和省略号"…"标明原文变更的地方。

That is, don't make the direct quotes too long. Otherwise it may distract readers from "the emphasis of the [key] words" (Hei and Guo 76).

如果文献作者是三位以上，则夹注格式为（第一位作者的姓氏+空格+et al.+空格+页码），文后文献著录可以把作者都列出来，也可以第一位作者的姓氏+空格+et al. 表示。例如：

Because of the annual increase of both the quantity and area of abandoned pits, great amounts of land resources are worsened, which leads to "serious contradiction among population growth, land utilization, and economic development" (Tan et al. 259).

四行或以上的引用则需另起一段，左边齐头缩进 10 个英文字符，不加引号。若引用段落本就是一自然段，则首行需再缩进 4~5 个字符。在引用段落的句号之后用圆括号标明作者姓氏及页码。例如：

The MLA documentation style is highly advised in the writing field of humanities, so just as Hei Yuqin and Guo Fenrong put it,

  It is suggested that you keep your working bibliography in this style at the very beginning. Having all the needed information like the titles, authors, dates, page numbers and URLs at your fingertips will save your frantic trips back to the library or the Internet. (Hei and Guo 35)

APA：用引号（""）标出引用部分，引号后紧跟着用圆括号夹注（Parenthetical Citation）的形式简要注明参考文献的出处，即（作者姓氏"last name"，发表年），必要时可加上页码，即（作者姓氏，发表年，页码），页码用"p. +空格+阿拉伯数字"表示。如作者在正文中已提及，就不再出现在括号夹注中。若引文有两个作者，其姓氏用 & 连接并标注于每次引用之后。例如：

It is "by the boundary conditions at the outermost grid points" that the general cubic splines and the natural cubic splines are different (Bulant & Klims, 2019, p. 138).

Bulant and Klims stress that it is "by the boundary conditions at the outermost grid points" that the general cubic splines and the natural cubic splines are different (2019, p. 138).

The author's own voice is of great value, "you can use your sources as support for your insights, not as the backbone of your paper" (Hei & Guo, 2017, p. 71).

Hei Yuqin and Guo Fenrong (2017) attach great value to the author's own voice, "you can use your sources as support for your insights, not as the backbone of your paper" (p. 71).

所引内容若不在一页上，则页码用"pp. +空格+阿拉伯数字"表示，如（Bulant & Klims, 2019, pp. 138-140）。

若 3~5 位作者，第一次引用时需列举全部的作者，往后若引用相同的文献，只需列出最主要作者的姓氏，再加上"et al."。文后参考文献部分，可把全部作者的姓名列举出来。如第一次引用：

For the success of the soil phytoremediation, "the identification of native species" is of great importance (Aman, Jafari, Reihan & Motesharezadeh, 2018).

再次引用：

It has been found out that the physical and chemical properties of the soil were "suitable for the cultivation of the seedlings" (Aman et al., 2018).

若作者 6 位以上，则列出第一位作者即可，格式应为"（作者 et al.，年份）"。同样，在参考文献部分，全部作者的姓名皆须列举出来。

The investigators find that several species of birds "that have been recorded by Sompud et al., 2013" can "be sighted again" in Gaya Island (Gilbert et al., 2018).

6.2.1.2 概括引用整篇文献的观点/改述

无论概括大意还是根据论文需要改述原文献部分内容，基本原则就是要与原文献相吻

6.2 英语学术论文参考文献引注规范简介及部分工程类英语期刊引注格式举例

合,不能断章取义或篡改原文献。MLA与APA在引用整篇文献观点或者改述原文献部分表达时,一般不用引号,出版年要有,页码标注不作硬性要求。但MLA与APA的具体标注情况仍有差异。需要指出的是,只要改述的内容在观点、句式结构、用词上与原文近似,即便有标注,亦有剽窃之嫌。因此,改述时要求作者在语言、句式结构上要有独创性。

当所引文献作者的姓氏未在正文中出现时:

MLA:放在引文后,标点符号前,格式为(作者姓氏 出版年)。如:

English linguistic imperialism appears especially in the countries where English is not a native language(Li 2019).

APA:放在引文后,标点符号前,格式为(作者姓氏,出版年)。如:

English linguistic imperialism appears especially in the countries where English is not a native language(Li, 2019).

当所引文献作者的姓氏已在正文同一句中出现时:

MLA:不需要使用夹注,可用 sb. claims…/sb. argues…/sb. points out…/sb. advocates…等引出所概括观点。

Li Huiying claims that English linguistic imperialism appears especially in the countries where English is not a native language.

APA:(出版年)放在引文后,标点符号前。作者姓氏不再重复。如出版年与姓氏均已在正文同一句话中出现,则不需使用夹注。

Li Huiying argues that English linguistic imperialism appears especially in the countries where English is not a native language(2019).

不是直接引用,页码的标注不作硬性要求。

6.2.1.3 在英语论文中引用中文著作或者期刊

MLA:(作者姓氏的汉语拼音+空格+页码),禁止使用汉字。例如:(Zhang 75),相应的参考文献格式举例如下:

Zhang Qingzong.[张庆宗],《外语学与教的心理学原理》.北京:外语教学与研究出版社,2009.

APA:括号夹注形式为(作者姓氏的汉语拼音,出版年)/(作者姓氏的汉语拼音,出版年,页码),禁止使用汉字。正文中的中文引文应提供英文译文,例如:

(Li, 2005)/(Li, 2005, p. 33).

6.2.1.4 非直接文献(Indirect Source/Secondary Source)的引用

为了保证文章的真实性、准确性、权威性以及对原作者的尊重,论文撰写过程中应尽量引用直接文献。如实在难寻直接文献,引文也可从非直接文献中析出,但要注意夹注及参考文献的格式。

MLA:

An investigator argued that the agency had been "strapped for money" (qtd. in Rybak 1A).

引用非直接文献以后,在著录正文后的参考文献时,只需列入该非直接文献的条目(即上述例子中的"Rybak"),参考文献格式与MLA格式吻合。

APA：

An investigator argued that the agency had been "strapped for money"（as cited in Rybak 1A）.

引用非直接文献以后,在著录正文后的参考文献时,只需列入该非直接文献的条目(即上述例子中的"Rybak"),参考文献格式与APA格式吻合。

6.2.1.5 同一作者多篇文献的引用

MLA：引用同一作者的多篇文献时,("作者姓氏 last name"+","+"作品名称"+"相关页码"),若作品名称过长,可使用其简短形式,如关键词,众所周知的简称,等。例如:

He thought it was his townsmen's misfortune " to have inherited farms, houses, barns, cattle, and farming tools"（Thoreau, Walden1）.

或者

Thoreau thought it was his townsmen's misfortune " to have inherited farms, houses, barns, cattle, and farming tools"（Walden1）.

或者

In his famous book Walden, Thoreau wrote that it was his townsmen's misfortune " to have inherited farms, houses, barns, cattle, and farming tools"（1）.

文后参考文献中,作者要求全名,表现形式为"作者姓氏,+名字+中间名的首字母.",如 Thoreau, Henry D.

APA：

夹注格式1：在不同的引用部分后分别加（作者姓氏,出版年）,不同的年份表示引用文献的发表次序。如:（Wang, 2001）（Wang, 2009）（Wang, 2018）。

夹注格式2：在最后一个引用部分后添加圆括号夹注,括号中包括作者姓氏及所有引用文献的出版年份,即（作者姓氏,出版年1,出版年2,出版年3）,出版年按照文章发表的先后顺序排序,不按照引用的先后排序。如:（Wang, 2001, 2009, 2018）。

同一年发表的文献,圆括号中在出版年后加上 a、b、c、⋯以示区别,参考文献中相应的条目里的年份亦加同样的字母与夹注相吻合（Smith, 2005a, 2005b）。

6.2.1.6 同时引用不同作者的多篇文献

括号夹注中按作者姓氏的字母先后顺序排列,不同作者用分号隔开。如:

MLA：

（David et al. 11-18; Tan et al. 259）

APA：

（David et al., 2016; Turner et al., 2002）

6.2.1.7 对引用文献资料的局部更改

在直接引用的过程中,为了使引用部分与文章内容相协调,避免语法错误,所引用文字会做局部更改,如对个别词汇进行删改或增减。无论删改还是添减字词均不能改变或歪曲所引语言的原意。MLA对此有更详细的规范。

删除引语中个别词句:用省略号（Ellipses）取代删除的词句。例如:

原文：whose misfortune it is to have inherited farms , houses, barns, cattle, and farming tools;

删改后：In his famous book Walden, Thoreau wrote that his townsmen were unlucky "…to have inherited farms, houses, barns, cattle, and farming tools"（Thoreau 1）

注意：如果是在句内删除某个词语，应该先空一格再加省略号；如果所删除内容是在一句整句之后，则应在该整句最后的标点（如句号、问号或感叹号）直接加省略号，不需空格。省略号应该用三个句点表示。

增添或者变更个别单词或词组：将需要增添或者变更的单词或词组放在方括号（square brackets）内。例如：

Thoreau（1）found that it was his townsmen's［whose］misfortune［it is］to have inherited farms.

6.2.2　MLA 与 APA 文后参考文献常见著录格式对比

文后参考文献著录在 MLA 规范里称为"Works Cited"，在 APA 规范里称为"References"。两种规范均要求文内引用要和文后的参考文献一致起来。

6.2.2.1　著录已出版的学术期刊文章（Printed Scholarly Journals）

（1）一位作者。

MLA：姓氏，名字."文章名."期刊名（斜体）卷.期（出版年）：页码.Print.

举例如下：

Gorman, Alice. "Ghosts in the Machine: Space Junk and the Future of Earth Orbit." *Architectural Design* 89.1（2019）：106-111. Print.

文章标题第一个词和冒号后第一个词的首字母必须大写；其余的词，除冠词、介词、并列连词以及不定式符号 to 以外，首字母必须大写。

APA：作者姓氏，名字首字母（大写）.（出版年）.文章名.期刊名（斜体），卷（期），页码.

举例如下：

Gorman, A.（2019）. Ghosts in the machine: Space junk and the future of earth orbit. *Architectural Design*, 89（1），106-111.

文章标题第一个词和冒号后第一个词的首字母必须大写，而其余的词，除专有名词以外，首字母均不需要大写。

如果作者有中间名，MLA 要求作者的姓名要完整，应标明首名的全称和中间名的首字母。APA 则要求作者的首名和中间名均用首字母表示，首字母大写。格式分别为：

MLA：姓氏，首名全称 中间名首字母（大写）.

如：Stewart, Donald C.

APA：姓氏，首名首字母（大写）.中间名首字母（大写）.

如：Stewart, D. C.

（2）两位作者。

MLA：姓氏 1，名字 1，and 名字 2 姓氏 2."文章名."期刊名（斜体）卷.期（出版年）：页码.Print.即只第一位作者的姓氏和名字颠倒，其他作者依然按照名在前姓在后的顺序。

Bulant, Petr, and Ludek Klims. "3-D Velocity Models - Transformation from General to Natural Splines. " *Studia Geophysica et Geodaetica* 63. (2019)：137-146. Print.

APA：姓氏1，名字首字母1（大写）.，& 姓氏2，名字首字母2.（出版年）.文章名.期刊名（斜体），卷（期），页码.

Bulant, P. , & Klims, L. (2019). 3-D velocity models-transformation from general to natural splines. *Studia Geophysica et Geodaetica*, 63, 137-146.

（3）三个或以上作者。

MLA：不同作者之间用逗号，倒数第一位和倒数第二位作者之间再加"and"。

格式一：姓氏1，首名1 中间名首字母1.，首名2 中间名首字母2. 姓氏2，and 首名3 中间名首字母3. 姓氏3. 即第一位作者的姓氏和名字颠倒，其他作者依然按照名在前姓在后的顺序。例如：

Booth, Wayne C. , Gregory G. Colomb, and Joseph M. Williams.

格式二：姓氏1，首名1 中间名首字母1. et al. 例如：

Booth, Wayne C. et al.

APA：不同作者之间用逗号，倒数第一位和倒数第二位作者之间再加"&"。格式如下：

姓氏1，首名首字母1. 中间名首字母1.，姓氏2. 首名首字母2. 中间名首字母2.，& 姓氏3. 首名首字母3. 中间名首字母3. 例如：

Booth, W. C. , Colomb. G. G. , & Williams. J. M.

6.2.2.2　著录已出版的书籍（Printed Books）

MLA：作者姓氏，名. 书名（斜体）（版本.）出版地：出版社，出版年. Print. 有两位作者的话，只需把第一位作者的姓氏在前名在后；第二位作者依旧按照名字在前姓氏在后的顺序。例如：

Coyle, William, and Joe Law. *Research Papers*. 16th ed. Beijing：Beijing Language and Culture University P, 2015. Print.

APA：作者姓氏，名.（出版年）. 书名（版本.）（斜体）. 出版地：出版社. 有两位作者的话，均按姓氏在前名在后的顺序，名字只写大写的首字母。例如：

Coyle, W. , & Law, J. (2015). *Research Papers* (16th ed.). Beijing：Beijing Language and Culture University Press.

6.2.2.3　著录网络文献

（1）网上学术期刊。

MLA：作者姓氏，名字. "文章名."期刊名（斜体）卷. 期（出版年）：页码.（若无页码可在此处写上 n. pag.）Web. 最新查看日期.

Lyster, Clare. "Disciplinary Hybrids：Retail Landscapes of the Post-Human City. " *Architectural Design* 89. 1 (2019)：100-105. Web. 12 June 2019.

MLA Handbook for Writers of Research Papers 7th Edition 认为，网址（URLs of Web Sources）过长，输入复杂且使用具有时效性，因此对于文献的著录价值不大。除非必须通过 URLs 才能找到文献，或者根据出版社或论文导师的要求必须添加网址，可把网址放到

查看日期之后。并且,如果网址过长必须换行的话,尽量在"//"或"/"处断开。例如:

Lyster, Clare. "Disciplinary Hybrids: Retail Landscapes of the Post-Human City. " *Architectural Design* 89. 1 (2019): 100–105. Web. 12 June 2019.

<http://specialsci.cn/detail/201901151609269555336824?view=download>.

APA:作者姓氏,名字首字母(大写). (出版年). 文章名. 期刊名(斜体),卷(期),页码. 文章网址.

Lyster, C. (2019). Disciplinary hybrids: Retail landscapes of the post-human city. *Architectural Design*, 89 (1), 100–105. Retrieved from http://specialsci.cn/detail/201901151609269555336824?view=download.

如果没有出版年,则把(出版年)改为(n. d.),为 no date 之意。根据网页提供信息不同,上述部分元素可以省略。

(2)纸印书籍的电子版(Online books with printed publication)。学术类书籍,为方便读者查询,一般都会有纸印版和电子版。

MLA:一般依次包括以下元素,其排列如下:作者. 书名(斜体). 版本. 出版地:出版社,出版年. 网站名(斜体). Web. 最新查询日期(日 月 年.) 根据实际情况,这些元素会有所取舍。举例如下:

Page, Scott E. *The Model Thinker: What You Need to Know to Make Data Work for You*. New York: Basic Books, 2018. *amazon. cn*. Web. 10 Sep. 2019.

Cascardi, Anthony J. *Ideologies of History in the Spanish Golden Age*. University Park: Pennsylvania State Up, 1997. *Penn State Romance Studies*. Web. 6 Nov. 2019.

APA:一般依次包括以下元素,其格式如下:作者. (出版年). 书名(斜体)[补充信息]. 网址(URLs of Web sources). 或者,如果有 DOI(即 Digital Object Identifier),也可用它来替代 URLs。DOI 是一个永久性的标识号,随着时间推移,网络书刊的 URLs 可能会有变化,但是 DOI 不会改变。格式如下:作者. (出版年). 书名(斜体)[补充信息]. doi:

Page, S. E. (2018). *The model thinker: What you need to know to make data work for you* [Basic Books version]. Retrieved from https://www.amazon.cn/dp/B07B8LHBVZ/ref.

Schiraldi, G. R. (2001). *The post-traumatic stress disorder sourcebook: A guide to healing, recovery, and growth* [Adobe Digital Editions version]. doi: 10. 1036/0071393722.

6.2.3 文后参考文献的排列

在 MLA 标准中,文后参考文献称为"Works Cited"。顾名思义,文后所列文献均在文内引用过,换言之,文内引用文献和文后著录文献要一致。"Works Cited"一般在文后另起一页,上页边距 1 英寸居中对齐。文献著录左对齐,左页边距 1 英寸。如果一条文献需占用两行或以上,则从第二行起,比第一行再缩进半英寸。各行之间都是双倍行距。

在 APA 标准中,文后参考文献称为"References",更强调文献的出版时间。要求双倍行距,悬挂缩进。同样,文后参考文献的著录只放入文内引用的来源,即文内引用文献和文后著录文献要一致。因此,一些文章才会把 Works Cited/References 和 Bibliography

（参考书目/文献目录）区分开来，前者是指文章引用的各种文献，包括书籍、期刊、报纸、杂志、音像资料、书信、网络资源等；后者是写作过程中参考的书目，是作者写作背景知识的来源，给作者以启迪，但不一定都在文中引用。对于个人交流内容，如书信、电子邮件、采访等，APA建议只在文内引用，不著录到文后的参考文献中。

（1）不同作者的文献。

MLA：按照作者姓氏的字母顺序依次排列。姓氏首字母相同，就看第二个字母的先后顺序，以此类推。如果多个作者姓氏完全相同，再按照名字的字母顺序依次排列。举例如下：

Booth, Wayne C.

Cascardi, Anthony J.

Colomb, Gregory G.

Coyle, William.

Joe Law.

Lyster, Clare.

Morris, Robert.

Morris, William.

Page, Scott E.

Pedro Orrego.

Peter Fleming.

APA：同样按照作者姓氏的字母顺序依次排列。其中需注意的是：

1）APA第六版181页提到"nothing precedes something"❶原则。比如：Clinton, D. 要排在 Clintonik, A. G. 之前，Novack, R. A. 要排在 Novacking, A. J. 之前。

2）Given name 在前、Sir name 在后的顺序适用于大部分西方国家，但在中国、日本、越南、新加坡、韩国等国家并不适用。同时，各出版社要求不一样，姓名的表达及前后顺序也不相同。比如汉语拼音 Wang Zhan，可以"Wang"是 Sir name，"Zhan"是 Given name，比如"王展"，也可以"Zhan"是 Sir name，"Wang"是 Given name，比如"詹望"。所以，在著录姓名时，一定要查找相关资料，确保其准确性，不要不加辨析，以讹传讹。

（2）同一作者（合著者）多个文献。

MLA：对于同一作者多个文献的著录，MLA规定，只在第一个条目中给出作者，以后的条目用"——"替代作者，后边仍用"."。但如果是编者、译者，则其后用"，"，并在书名前加编、译等的缩写（ed., trans., comp.）。条目排列顺序按照书名的字母排序，但是如果书名第一个单词是"a, an, the"，则直接略过，按照后边词的字母排序。例如：

Clinton, David. *Linux in Action*. Greenwich：Manning Publications, 2018. Print.

——. *Practical Lpic-1 Linux Certification Study Guide*. Berkeley：Apress, 2016. Print.

Mark Twain. [Clemens, Samuel L.]. *The Adventures of Tom Sawyer*. New York：Penguin

---

❶ American Psychological Association. Publication manual of the American Psychological Association [M]. 6th ed. Washington DC：American Psychological Association, 2010.

Random House, 1995. Print.

——. *The Prince and the Pauper*. New York: Penguin Random House, 2002. Print.

MLA 标准中，如果所引文献既有笔名/化名（Pseudonym），又有真实姓名，标注方式有两种：1）按照更为大众所知的名字标注，比如上述例子中只标注 Mark Twain（同时因为 Mark Twain 是笔名，原是密西西比河水手使用的表示在航道上所测水的深度的术语，"两个标记"，亦即水深两浔。不是普通的姓名，所以此处只按照原格式著录 Mark Twain）。2）在笔名后用中括号标注真实姓名。不同条目中，合著者的第一作者相同，则按照第二作者的姓氏字母排序；第二作者相同，则按照第三作者姓氏字母排序，以此类推。（注意：MLA 格式要求，两位或以上作者时，除了第一作者是按照先姓氏后名字的顺序，其余作者均按照习惯先名字后姓氏排序。）

Coyle, John J., C. John Langley, Jr., Robert A. Novack, and Brian J. Gibson. *Managing Supply Chains: A Logistics Approach*. 9th ed. Beijing: Publishing House of Electronics Industry, 2016. Print.

Coyle, John J., Robert A. Novack, and Brian J. Gibson. *Transportation-A Global Supply Chain Perspective*. 9th ed. Stanford: Cengage Learning, 2018. Print.

APA：此处 APA 规定更为详细。

1）一位作者（或合著者）多个文献，在著录时每个条目都要包括作者，条目按照出版时间先后排序。

Clinton, D. (2016). *Practical lpic-1 linux certification study guide*. Berkeley: Apess.

Clinton, D. (2018). *linux in action*. Greenwich: Manning Publications.

2）独立作者同时又是合著文献的第一作者：独立作者作品在前，合著在后（不考虑发表时间）。

Coyle, J. J. (2019).

Coyle, J. J., Novack, R. A. & Gibson B. J. (2018).

3）合著文献中只有第一作者相同：按照第二作者姓氏排序，若第二作者姓氏也相同，按第三作者姓氏排序，以此类推。

Coyly, M., Langley, J., Novack, R. A. & Gibson B. (2015).

Coyly, M., Novack, R. A. & Gibson J. (2013).

4）不同条目的多位作者，只有第一作者的姓氏相同：按照姓氏后边第一个大写的名字首字母排序。

Coyle, D. L., & Lawyer, M. (2013).

Coyle, W., & Law, J. (2015).

6.2.4 部分工程类英语期刊引用格式举例

本部分以几个工程类英语期刊为例，进一步阐明文内引用的格式。为了方便读者对该期刊引用格式有较全面的了解，同时列举了相关的参考文献部分。

6.2.4.1 文内引用及文后参考文献著录完全按照 APA/MLA 格式

Journal of Industrial Ecology 期刊中文名《工业生态学杂志》，由美国发行，Wiley-Blackwell 出版，是 SCIE 来源期刊，季刊。据 2019 年 12 月最新版本的中科院 SCI 期刊分区，它属于大类学科中的环境科学与生态学 2 区，小类学科中的工程环境、绿色可持续发展技术 3 区，环境科学 2 区。

下面以论文 Working Conditions in Hydrogen Production: A Social Life Cycle Assessment❶为例，看 Journal of Industrial Ecology 对文内引用及文后参考文献著录的要求。部分内容及文后参考文献如图 6-1 和图 6-2 所示。

> The SDGs were also integrated into the work of the International Labour Organization (ILO) to derive key areas for working conditions that need to be addressed in order to ensure Sustainable Development (ILO, 2016). As indicator selection for SLCA is still a contentious issue, this case study uses the six ILO key areas (Table 1) as a basis (Hunkeler, 2006; Kühnen & Hahn, 2017; Lehmann, Zschieschang, Traverso, Finkbeiner, & Schebek, 2013). Table 1 further illustrates the connection between the key areas and the impact categories in PSILCA 2.0.
>
> In order to quantify the social impact along the AEL hydrogen production chain in Germany, Austria, and Spain as well as their upstream supply chains, the PSILCA 2.0 database is used (Eisfeldt & Ciroth, 2017). The functional unit is 1 kg of hydrogen produced. The database is based on a multiregional input/output database called Eora (Lenzen, Kanemoto, Moran, & Geschke, 2012; Lenzen, Moran, Kanemoto, & Geschke, 2013), which enables a global mapping of supply chains on the sectoral scale, following the idea of life cycles. Information on social and socioeconomic issues is collected from global statistical agencies and internationally recognized organizations, for example, World Bank, ILO, WHO. The raw data

图 6-1　Journal of Industrial Ecology 文内引用举例

> **REFERENCES**
>
> Aaronson, S. (2010). How China's employment problems became trade problems? *Global Economy Journal, 10*(3), 31.
>
> Anker, R., & Anker, M. (2013). *A shared approach to estimating living wages—Short description of the agreed methodology*. London: The Global Living Wage Coalition, ISEAL Alliance.
>
> Benoît Norris, C., Cavan, D. A., & Norris, G. (2012). Identifying social impacts in product supply chains: Overview and application of the social hotspot database. *Sustainability, 4*(9), 1946–1965.
>
> Benoît Norris, C., & Norris, G. (2015). The social hotspots database context of the SHDB. In J. Murray, D. McBain, & T. Wiedmann (Eds.), *The sustainability practitioner's guide to social analysis and assessment* (pp. 52–73). Champaign, IL: Common Ground Publishing LLC.
>
> Corona, B., Bozhilova-Kisheva, K. P., Olsen, S. I., & San Miguel, G. (2017). Social life cycle assessment of a concentrated solar power plant in Spain: A methodological proposal. *Journal of Industrial Ecology, 21*(6), 1566–1577.
>
> Chan, C. K.-C., & Hui, E. S.-I. (2014). The development of collective bargaining in China: From "Collective bargaining by riot" to "Party state-led wage bargaining." *The China Quarterly, 217*, 221–242.
>
> Christofides, L. N., Polycarpou, A., & Vrachimis, K. (2013). Gender wage gaps, 'sticky floors' and 'glass ceilings' in Europe. *Labour Economics, 21*, 86–102.
>
> Dreyer, L., Hauschild, M., & Schierbeck, J. (2006). A framework for social life cycle impact assessment (10 pp). *The International Journal of Life Cycle Assessment, 11*(2), 88–97.
>
> EC. (2018). The gender pay gap situation in the EU: Causes of the gender pay gap in the EU, the pay gap situation in different EU countries. Retrieved from https://ec.europa.eu/info/policies/justice-and-fundamental-rights/gender-equality/equal-pay/gender-pay-gap-situation-eu_en#differencesbetweeneucountries
>
> Eisfeldt, F., & Ciroth, A. (2017). PSILCA—A Product Social Impact Life Cycle Assessment database. Database version 2. Berlin: GreenDelta GmbH.
>
> Foubert, P. (2017). *The enforcement of the principle of equal pay for equal work or work of equal value*. Brussels, Belgium: European Equality Law Network (European Commission).

图 6-2　Journal of Industrial Ecology 文后参考文献著录举例

从图 6-1 和图 6-2 来看，本文严格按照 APA 的文献引注格式。首先，从文内引用来看，当概括引用来自独立作者或组织的文献时，采用了（作者，出版年）格式，如（ILO, 2016）；当有两个作者时，用（姓氏1 & 姓氏2, 出版年）格式，如（Eisfeldt & Ciroth, 2017）；当同时引用多个作者的多个文献时，用（作者姓氏1, 出版年1; 作者姓氏2, 出版年2; 作者姓氏3, 出版年3）的格式，而且当一部文献的作者为三者或以上时，

---

❶　Werker J, Wulf C, Zapp P. Working conditions in hydrogen production: a social life cycle assessment [J]. Journal of Industrial Ecology, 2019, 1: 10.

采用（作者姓氏1，作者姓氏2，作者姓氏3，& 作者姓氏4，出版年）格式，如（Lenzen, Kanemoto, Moran, & Geschke, 2012; Lenzen, Moran, Kanemoto, & Geschke, 2013）。其次，从文后文献著录来看，参考文献与文内引用部分相吻合，文献条目按照作者姓氏字母顺序排列。著录已出版的书籍，按照"作者姓氏，名.（出版年）.书名（版本）.出版地：出版社."格式，比如：

Anker, R., & Anker, M.（2013）. A shared approach to estimating living wages—Short description of the agreed methodology. London：The Global Living Wage Coalition, ISEAL Alliance.

著录已出版期刊，按照"作者姓氏，名字首字母（大写）.（出版年）.文章名.期刊名（斜体），卷（期），页码."的格式，文章标题第一个词和冒号后第一个词的首字母大写，而其余的词，除专有名词以外，首字母均不大写。作者有中间名时，作者的首名和中间名均用首字母表示，首字母大写。比如：

Aaronson, S.（2010）. How China's employment problems became trade problems? *Global Economy Journal*, 10（3），31.

还有网络资源的著录，比如：

EC.（2018）. The gender pay gap situation in the EU：Causes of the gender pay gap in the EU, the pay gap situation in different EU countries. Retrieved from https：//ec.europa.eu/info/policies/justice-and-fundamental-rights/gender-equality/equal-pay/gender-pay-gap-situation-eu_en#differences between EU countries

同时还出现了引用多个作者多个文献时第一作者相同的情况，比如：

Lenzen, M., Kanemoto, K., Moran, D., & Geschke, A.（2012）. Mapping the structure of the world economy. *Environmental Science & Technology*, 46（15），8374-8381.

Lenzen, M., Moran, D., Kanemoto, K., & Geschke, A.（2013）. Building Eora：A global multi-region input-output database at high country and sector resolution. *Economic Systems Research*, 25（1），20-49.

APA格式常见于国外的心理学、教育学及自然科学类期刊，比如Journal of Industrial Ecology和Journal of Operations Management。但是，实际操作过程中，一些土木、测绘、建筑类英文期刊，稿件格式在APA/MLA基础上会有所变化。

6.2.4.2　文内引用及文后参考文献著录在APA/MLA格式基础上的变化

（1）Earth, Planets and Space期刊中文名《地球，行星和宇宙空间》，由日本发行、德国Springer出版，是科学类SCI期刊。根据2019年12月最新版本的中科院SCI期刊分区，它属于地球科学综合3区。

下面以论文The Effects of Dislocations on Crystallographic Twins and Domain Wall Motion in Magnetite at the Verwey Transition❶为例，看Earth, Planets and Space对文内引用的要求。部分内容及文后参考文献如图6-3所示。

其文内引用采用夹注的形式，即在正文中插入圆括号，标注引文的文献来源。文中夹

---

❶ Lindquist et al. The effects of dislocations on crystallographic twins and domain wall motion in magnetite at the verwey transition [J]. Earth, Planets and Space, 2019, 71：5.

> At ~ 120 K, magnetite undergoes a first-order phase transition, called the Verwey transition (Li 1932; Verwey and Haayman 1941). At the Verwey temperature (T_V), the crystallographic structure of magnetite switches from cubic ($Fd\bar{3}m$) to monoclinic (Cc) (Kasama et al. 2013; Iizumi et al. 1982; Senn et al. 2012). This transformation is accomplished via slight, predictable, and reversible shifts in atomic positions resulting in redefined crystallographic axes such that the new monoclinic a, b, and c-axes correspond to the cubic [1 $\bar{1}$ 0], [110], and [001] axes, respectively (Abe et al. 1976; Iida 1980). The magnetic easy axis changes from the cubic ⟨111⟩ directions to this new monoclinic [001] (c-axis), and the

> **References**
> Abe K, Miyamoto Y, Chikazumi S (1976) Magnetocrystalline anisotropy of low temperature phase of magnetite. J Phys Soc Jpn 41(6):1894–1902
> Aragon R, Buttrey DJ, Shepherd JP, Honig JM (1985) Influence of nonstoichiometry on the Verwey transition. Phys Rev B 31(1):431–436
> Bezaeva N, Swanson-Hysell N, Tikoo S, Badyukov D, Kars M, Egli R, Chareev D, Fairchild L, Khakhalova E, Strauss B et al (2016) The effects of 10 to >160 GPa shock on the magnetic properties of basalt and diabase. Geochem Geophys Geosyst 17(11):4753–4771
> Bickford LR (1950) Ferromagnetic resonance absorption in magnetite single crystals. Phys Rev 78:449–457
> Bickford LR (1953) The low temperature transformation in ferrites. Rev Mod Phys 25:75–79
> Brabers VAM, Walz F, Kronmüller H (1998) Impurity effects upon the Verwey transition in magnetite. Phys Rev B 58(21):14163–14166
> Bryson JFJ, Kasama T, Dunin-Borkowsi RE, Harrison RJ (2013) Ferrimagnetic/ferroelastic domain interactions in magnetite below the Verwey transition: part II. Micromagnetic and image simulations. Phase Transit 86(1):88–102
> Calhoun BA (1954) Magnetic and electric properties of magnetite at low

图 6-3　期刊 Earth，Planets and Space 文内引用及部分参考文献举例

注的信息来源于正文后面的参考文献。从文内引用来看，为典型的 MLA 格式：作者和出版年之间为空格，如（Li 1932），两个作者用"and"连接，如（Verwey and Haayman 1941），三个或以上作者夹注格式为第一位作者的姓氏+空格+et al.+空格+页码，如（Abe et al. 1976），同时引用不同作者的多篇文献时，括号夹注中按作者姓氏的字母先后顺序排列，不同作者用分号隔开，如（Li 1932；Verwey and Haayman 1941）。所引文献出处均以夹注的形式加以注释。但是，文后的参考文献著录部分用的"References"而不是"Works Cited"，著录规则基本上按照 APA 标准（即"作者姓氏，名字首字母（大写）.（出版年）. 文章名. 期刊名（斜体），卷（期），页码."），参考文献也按作者姓氏的字母排序。但是在作者姓名的著录上与 APA 标准有出入，比如 Abe K，Miyamoto Y，Chikazumi S（1976）应为 Abe K., Miyamoto Y., & Chikazumi S.（1976）.；Bickford LR（1950）应为 Bickford L. R.（1950）.

（2）Tunneling and Underground Space Technology 期刊中文名《隧道与地下空间技术》，出版地英国，出版商为 Elsevier Ltd，双月刊，SCI、EI 收录期刊。根据 2019 年 12 月最新版本的中科院 SCI 期刊分区，它属于小类学科中工程土木类 2 区。

下面以论文 Reclamation and Reuse of Abandoned Quarry：A Case Study of Ice World & Water Park in Changsha❶ 为例，看 Tunneling and Underground Space Technology 对文内引用的要求。部分内容及文后参考文献如图 6-4 所示。

其文内引用亦采用夹注的形式，但将 APA 格式和 MLA 格式混合在一起。比如 MLA 格式要求，所引文献有三个或三个以上作者，第一次引用时夹注格式为第一位作者的姓氏+空格+et al.+空格+出版年/页码；但在 APA 格式中，若 3~5 位作者，第一次引用时需列举全部的作者，往后若引用相同的文献，只需列出最主要作者的姓氏，再加上"et al."，作者和出版年之间用逗号隔开。就本文来看，第一次引用就直接用了"第一位作者的姓氏+

---

❶ Tan et al. Reclamation and reuse of abandoned quarry：a case study of ice world & water park in Changsha [J]. Tunneling and Underground Space Technology，2019，85：259~267.

> the ecological restoration of the open mining pits in Lausitze of Germany (Krümmelbein et al., 2012). These cases have drawn wide public attention, resulted in an integrated and diversified way of reuse of abandoned mines thereafter.
>
> The abandoned pits can be reused through many ways, such as agricultural application, resource storage, underground warehouses, and tourism development (Peila and Pelizza, 1995). The abandoned pits in Heerlen, Limburg Province, Netherlands were turned into power stations for utilizing the geothermal resources (Roijen and Op't Veld, 2007). The abandoned tunnels of Wieliczka Salt Mine, which was located in the suburb of Krakow City in Poland, were rebuilt into an ancient salt mine museum for exhibiting the relics of salt industry and salt carving works, and as well as for the treatment of respiratory diseases (Krakowiak, 2013). An abandoned gold mine in Blake, South Dakota, America was turned into a deep underground laboratory to create a deep underground environment for the experiments in the frontier of particle physics (Cowen, 2010). In some other cases, forests, grasslands, and fishponds have been reclaimed from abandoned mines through the backfilling, stripping, and covering of soil for agricultural cultivation (Tokgöz, 2013; Otchere et al., 2004). Currently, most abandoned mines are reused in this way in China.
>
> Transforming the pits into such places as parks and museums by adding recreational elements, the tourism development of abandoned
>
> Germano, D., Machado, R., Godinho, S., Santos, P., 2016. The impact of abandoned/disused marble quarries on avifauna in the anticline of Estremoz, Portugal: does quarrying add to landscape biodiversity? Landsc. Res. 41 (8), 880-891.
> Jones, A.C., Hamilton, D., Purvis, M., Jones, M., 2001. Eden project, Cornwall: design, development and construction. Struct. Eng. 79 (20), 30-36.
> Jordán, M.M., García-Sánchez, E., Almendro-Candel, M.B., Navarro-Pedreño, J., Gómez-Lucas, I., Melendez, I., 2009. Geological and environmental implications in the reclamation of limestone quarries in Sierra de Callosa (Alicante, Spain). Environ. Earth Sci. 59 (3), 687.
> Krakowiak, B., 2013. Museums in cultural tourism in Poland. Turyzm 23 (2), 23-32.
> Krümmelbein, J., Bens, O., Raab, T., Naeth, M.A., 2012. A history of lignite coal mining and reclamation practices in Lusatia, eastern Germany. Can. J. Soil Sci. 92 (1), 53-66.
> Meech, J.A., McPhie, M., Clausen, K., Simpson, Y., Lang, B., Campbell, E., Condon, P., 2006. Transformation of a derelict mine site into a sustainable community: the Britannia project. J. Clean. Prod. 14 (3), 349-365.
> Meyer, E.K., 1991. The public park as Avante-Garde (Landscape) architecture: a comparative interpretation of two parisian parks, Parc de la Villette (1983-1990) and Parc des Buttes-Chaumont (1864-1867). Landsc. J. 10 (1), 16-26.
> Otchere, F.A., Veiga, M.M., Hinton, J.J., Farias, R.A., Hamaguchi, R., 2004. Transforming open mining pits into fish farms: moving towards sustainability. Nat. Resour. Forum 28 (3), 216-223.
> Peila, D., Pelizza, S., 1995. Civil reuses of underground mine openings: a summary of international experience. Tunn. Undergr. Sp. Technol. 10 (2), 179-191.
> Preston, D., 2003. The Story of Butchart. Gardens. Highline Publishing.
> Tokgöz, N., 2013. Use of TBM excavated materials as rock filling material in an abandoned quarry pit designed for water storage. Eng. Geol. 153, 152-162.
> Worthington, K., Bolenbaugh, S., 2008. From landfill to park: stormwater runoff as a benefit, not a burden. Proc. Water Environ. Feder. 16, 1027-1034.
> Zhang, R., 2012. China: Luxury Hotel 'groundscraper' Planned in Abandoned Quarry. CNN Travel.

图 6-4　期刊 Tunneling and Underground Space Technology 文内引用及部分参考文献举例

空格+et al. +逗号+出版年或页码", 如 (Krümmelbein et al., 2012), 但逗号的运用又使它区别于纯粹的 MLA 格式。APA 格式中两个作者则用 "&" 连接, 如 (Verwey & Haayman, 1941), 但本文相关引用明显融合了 MLA 和 APA 的特点, 如 (Peila and Pelizza, 1995), 逗号的运用使它区别于纯粹的 MLA 格式, 用 "and" 代替 "&" 又使它区别于纯粹的 APA 格式。从参考文献著录来看, Tunneling and Underground Space Technology 的文章, 文后 "References" 以作者姓氏字母排序, 但是作者姓名、出版年及杂志的标注格式明显有改动的痕迹。

所以, 尽管 MLA 和 APA 格式在英文期刊中比较普遍, 但因各杂志要求不同, 投稿时一定要提前关注期刊对文献引用的格式要求, 不要因为格式原因影响审稿。

(3) Indoor and Built Environment 期刊中文名《室内与建筑环境》, 出版地英国, 出版商为 SAGE Publications Ltd, 双月刊, SCI 收录期刊。根据 2019 年 12 月最新版本的中科院 SCI 期刊分区, 它属于大类学科中工程技术 4 区, 小类学科中结构与建筑技术 4 区, 工程环境 4 区。

下面以论文 Effects of Shaft Inclination Angle on the Capacity of Smoke Exhaust under Tunnel Fire❶ 为例, 看 Indoor and Built Environment 对文内引用的要求。部分内容及文后参考文献如图 6-5 和图 6-6 所示。

---

❶ Yao, et al. Effects of shaft inclination angle on the capacity of smoke exhaust under tunnel fire [J]. Indoor and Built Environment, 2019, 28: 77~87.

Effects of shaft inclination angle on the capacity of smoke exhaust under tunnel fire

Yongzheng Yao[1], Shaogang Zhang[1], Long Shi[2] and Xudong Cheng[1]

Abstract
Vertical shaft is one of the most important approaches for smoke control under tunnel fires. However, the boundary layer separation is a common phenomenon of hampering the smoke exhaust for vertical shafts. A tilted shaft has been proposed to solve problems and improve the capacity of smoke exhaust. In this study, the effect of shaft inclination angle (θ decreases from 90° to 14°) and shaft height on the capacity of smoke exhaust was addressed numerically. A series of scenarios were simulated in a full-scale road tunnel. Numerical results showed that the tilted shaft could eliminate the boundary layer separation. However, small shaft inclination angle could lead to a relatively higher resistance to the smoke and a smaller cross-section area of shaft, which could have an adverse effect on the capacity. Under these two factors, an optimal inclination angle exists in the shaft of around 76° in this study. Based on the smoke flow characteristics and exhaust effect, the inclination angle was roughly divided into three regions. The main influence factor of the inclination angle on the mass flow rate of smoke in each region was examined. For a comprehensive consideration, the low and slightly tilted shaft was applied to tunnel fires, which can improve the capacity of smoke exhaust obviously.

Keywords
Tunnel fire, Vertical shaft, Smoke exhaust, Boundary layer separation, Tilted shaft

Accepted: 11 September 2017

Introduction

Road tunnels as an effective method to relieve the traffic congestion and these have been built widely around the world.[1] Due to its special long-narrow structure,[2,3] tunnel fires are always catastrophic with fast-spreading poisonous smoke and heat.[4–6] For example, the Daegu Subway Station Fire in South Korea happened in 2003 killed 198 people.[2] The Baku railway tunnel fire in Azerbaijan happened in 1995 in which 289 people were killed and 265 were injured[4] and the Mont-Blanc road tunnel fire in 1999 in France bordering with Italy where 39 people were killed.[5] According to previous studies,[7–9] about 75–85% of deaths were because of the poisonous smoke under fire conditions. In addition, hot smoke released from fire sources not only would affect passengers' health and mortality in the tunnel, but could also damage the tunnel structures;[10] leading to secondary damages when the tunnel being collapsed under high temperature. Hence, exhausting the hot smoke in the tunnel is of great importance to reduce safety risks to structures and people's health.

Vertical shaft, as an effective approach for smoke exhaust in tunnel fires, has attracted increasing attention from both engineers and relevant researchers. Fan et al.[11] pointed out that the mechanical ventilation would require a large space of excavation for installation of ventilation equipment, but the natural

[1]State Key Laboratory of Fire Science, University of Science and Technology of China, Hefei, PR China
[2]Civil and Infrastructure Engineering Discipline, School of Engineering, RMIT University, Melbourne, Australia

Corresponding author:
Xudong Cheng, State Key Laboratory of Fire Science, University of Science and Technology of China, Jinzhai Road 96#, Hefei 230027, China.
Email: chengxd@ustc.edu.cn

```
References
1. Fan CG and Yang J. Experimental study on thermal smoke
   backlayering length with an impinging flame under the tunnel
   ceiling. Exp Thermal Fluid Sci 2017; 82: 262–268.
2. Yao YZ, Cheng XD, Zhang SG, Zhu K, Zhang HP and Shi L.
   Maximum smoke temperature beneath the ceiling in an enclosed
   channel with different fire locations. Appl Therm Eng 2017; 111:
   30–38.
3. Wu Y, Gao NP, Wang LH and Wu XP. A numerical analysis of
   airflows caused by train-motion and performance evaluation of a
   subway ventilation system. Indoor Built Environ 2014; 23:
   854–863.
4. Nilsen AR and Log T. Results from three models compared to
   full-scale tunnel fires tests. Fire Saf J 2009; 44: 33–49.
5. Vuilleumier F, Weatherill A and Crausaz B. Safety aspects of
   railway and road tunnel: example of the Lötschberg railway
   tunnel and Mont-Blanc road tunnel. Tunn Undergr Space
   Technol 2002; 17: 153–158.
6. Hu LH, Huo R and Chow WK. Studies on buoyancy-driven
   back-layering flow in tunnel fires. Exp Thermal Fluid Sci 2008;
   32: 1468–1483.
7. Yang D, Huo R, Zhang XL and Zhao XY. Comparison of the
   distribution of carbon monoxide concentration and temperature
   rise in channel fires: reduced-scale experiments. Appl Therm Eng
   2011; 31: 528–536.
8. Alarie Y. Toxicity of fire smoke. Crit Rev Toxicol 2002; 32:
   259–289.
9. Hietaniemi J, Kallonen R and Mikkola E. Burning characteris-
   tics of selected substances: production of heat, smoke and chem-
   ical species. Fire Mater 1999; 23: 171–185.
```

图 6-6 期刊 Indoor and Built Environment 文后参考文献举例

如图 6-5 和图 6-6 所示，期刊 Indoor and Built Environment 文内注释分两部分：首先，作者简介（单位等）以脚注形式体现。脚注，又称页末注、页下注、呼应注。作为对文档中某个内容的注释或对引用的补充说明，在所要加注处的右上角标注数码，同时在本页页面最底端留出位置，划一横线与正文隔开，在横线下用同样数码标注注释内容。多应用于篇幅较短、引用文献较少的文章。对于文内引用的注释，一般包括作者姓名、书名、出版地、出版社、出版时间及页码。不同出版社对脚注的标注符号要求也有所不同。此外，期刊杂志中，收稿日期、基金项目、作者简介多以脚注的形式表现出来。期刊 Indoor and Built Environment 中用简单的阿拉伯数字 1、2、3、…右上标，在首页末同样用 1、2、3 依次标明作者单位及地址。其次，正文的引用部分，采用附注形式。附注是指在正文需注释处的右上方按顺序加注数码，即右上标。在正文之后写明"附注"或"注释"字样，然后依次写出数码和注解。在实际操作中，根据出版社或各高校的要求不同，数码的表达形式可为简单的 1、2、3、…（如期刊 Indoor and Built Environment），也可为［1］、［2］、［3］、…（如国内的核心期刊土木工程学报、山东建筑大学学报），或①、②、③、…。正文后的附注或注释也可直接以参考文献的形式体现出来（如图 6-6 所示）。

6.3 引用过程中的预防剽窃

6.3.1 什么是剽窃

剽窃（Plagiarism），就是把他人的作品或者观点据为己用却不注明出处，或有意把他人的作品或观点视作己出，不给出恰当的引用从而让读者误以为是作者原创。Richard Allen Posner 在他的 The Little Book of Plagiarism 指出，"Plagiarism…consists of unauthorized copying that the copier claims (whether explicitly or implicitly, and whether deliberately or carelessly) is original with him and the claim causes the copier's audience to behave otherwise than it would if it knew the truth"❶。

剽窃分为有意识剽窃和无意识剽窃。有意识剽窃是主观上有剽窃的意图，故意把他人学术成果收为己用，混淆视听，以求快速出成果。这是造成学术不端的主要原因。2016年9月1日，《高等学校预防与处理学术不端行为办法》正式实施，从官方角度对学术不端现象进行整治。

无意识剽窃是指由于对待所引内容的态度不严谨，或对引用规则的不了解而造成的对他人已有学术成果的侵犯，比如部分引用原文却没用引号标注，把原作者姓名拼错、漏写或把姓氏与名字混淆，引用网络资源不加注释，过度间接引用而对其正确性不加辨析以致以讹传讹恶化学术环境等。

6.3.2 如何在引用过程中预防剽窃

（1）端正学术态度。思想意识引领行动，决定行动方向。做学问，必须要有端正的学术态度。踏踏实实做学问，杜绝抄袭。要从主观上避免抄袭行为，不抱有侥幸心理，不急功近利，只要不是原创，一定要注明所引内容出处，直接引用要用引号标出。

（2）了解引用规则，有意规避无意识的剽窃。了解引用规则的目的是为了避免由于粗心或对规则的不熟悉而无意识地"剽窃"他人成果。MLA 格式和 APA 格式都有明确的引用规则，各期刊杂志出版社也有具体的投稿要求，写文之前一定要明确引用规则以避免不必要的麻烦。

实际上，在搜集整理相关资料记笔记的过程中就要有意识避免剽窃。记笔记的方式可以分为高度概括总结、改述或直接引用已有学术内容或观点。总结时要用独创性语言总结引文大意或观点，用词简单明了，句式结构和语言上要区别于原文，但要做到"信"于原文。如若引用原文中的词汇，一定用引号且注明出处。而改述则同样要求引用者自己的原创语言及句式结构。黑玉琴，郭粉绒在《英语学术论文写作（本科用）》提到"literal paraphrasing"和"free paraphrasing"❷。前者是指逐字逐句地利用同义词或近义表达来复述引用部分，但这样容易因为词性不同搭配不同而导致句式结构不符合英语习惯，从而造成表达错误或语义混乱；后者是指用自己的语言和句式结构来复述，但语义与原文必须一致。这要求改写者对原文的高度理解和对词汇的严格把控，才能做到"信"和"达"。而对于直接引用，一定要注意所引内容要在意义及创新性方面不可替代，且语言生动有说服力，或者能强烈支撑引用者自己的文章或观点，而不是为了引用而引用、随意引用任意词。

（3）需要规避的几点。

1）引文和文后的参考文献不一致。原因之一，撰文过程中无意识漏掉参考文献，引

---

❶ Posner R A. The Little Book of Plagiarism [M]. New York：Pantheon，2007：106.
❷ 黑玉琴，郭粉绒. 英语学术论文写作（本科用）[M]. 西安：西安交通大学出版社，2017.

文和参考文献不统一；原因之二，作者有意识增加参考文献以彰显阅读量大、知识储备丰足。对于前者，写文过程中一定要有足够的耐心和细心，涉及引用之处，立刻找到引文出处并保存。养成及时存盘的好习惯，并在文章成稿后仔细核对文内引用和文后参考文献。对于后者，要求作者要有脚踏实地的学术精神，对读者、对自己的作品负责。提高了不相关作品的引用率，对其他引文、作者有失公允。

2）把间接引用当作直接引用。在撰文过程中，将未查阅过原始出处的引文作为直接引用，将引自译著的引文或作者自己的译文标注为引自原著。这样不加甄别，以讹传讹，学术环境终将被恶化。

3）模糊引用。即在文中不标明出处，引文混在文内不加标注，但在参考文献部分却列出，试图推却剽窃的嫌疑。

6.4 范文阅读与分析

6.4.1 范文阅读

以下导入部分摘自 Emily A. Gilbert, Jephte Sompud, Oswald A. Igau, Maklarin Lakim, Rimi Repin, Alim Biun 2018 年发表在 Transactions on Science and Technology 的文章 An Update on the Bird Population in Gaya Island[①]。

Introduction

Islands are particularly small in size as opposed to large habitat or area such as a continent. It also has its' own functional ecosystem within the small confined areas (Taylor & Kumar, 2016). The unique geological landscape of an island does not only offer mesmerizing beauty of the island's scenery but harbors both marine and terrestrial wildlife (Rodrigues & Cunha, 2012). Apart from that, small islands are well known for harboring species of animals and plants that have high level of endemism (Kier et al., 2009). Past studies have shown that islands are also a vital site in providing habitat for birds especially the endemic, migratory (Turner et al., 2002; David et al., 2016) as well as threatened species (Rodrigues & Cunha, 2012). Birds have been widely used by researchers as an effective biological indicator (Sodhi et al., 2005) for its' ability to respond quickly the changes of the environment (Yap et al., 2007). Previous studies have shown that bird also plays crucial roles in the ecosystem such as the seed disperser, pollinator (Peh et al., 2005), as well as predators in the food chain (Basnet et al., 2016) across different landscapes around the world including the small islands. In Sabah, there are still limited avian studies that have been conducted in small forested islands to provide fundamental understanding of the bird community that inhabit in an isolated and confined ecosystem such as Gaya Island. To date, the published study about the avian in Gaya island is only limited to Well 1976, Sompud et al.,

---

① Gilbert, et al. An update on the bird population in Gaya island [J]. Transactions on Science and Technology, 2018, 5 (2): 171~176.

2013 and Sompud et al., 2016. Hence, this study aims to document and update the species composition of the bird population in Gaya Island of Sabah.

6.4.2 范文分析

从文内引用的角度看，发表在 Transactions on Science and Technology 的文章多采用 APA 格式（注意作者和出版年之间用逗号连接）。节选部分出现了同一著作两个作者的情况（注意是作者姓氏用 & 连接，姓氏与出版年之间用逗号隔开），如（Taylor & Kumar, 2016），（Rodrigues & Cunha, 2012），也出现了三个及以上作者，如（Kier et al., 2009），（Peh et al., 2005），同时也有对不同作者不同文章的概括引用（两个不同作者文章用分号隔开），如（Turner et al., 2002; David et al., 2016），但是，很显然此处没有按照姓氏首字母排序。总之，所摘录文献绝大部分标注规范，但或许作者是从查重率的角度出发，整篇文章中没有直接引用部分。

为方便读者对该格式的全面理解，文内引用对应的参考文献摘录如下（参考文献是按姓氏字母顺序排序，前边的数字是它们在原参考文献中的序号）：

［1］Basnet, T. B., Rokaya, M. B., Bhattarai, B. P. & Münzbergová, Z.(2016). Heterogeneous Landscapes on Steep Slopes at Low Altitudes as Hotspots of Bird Diversity in a Hilly Region of Nepal in the Central Himalayas. PLoS ONE, 11 (3), 1-19.

［4］David, G., Roslan, A., Mamat, M. A., Abdullah, M. T. & Hamza, A. A. (2016). A Brief Survey on Birds from Pulau Perhentian Besar, Terengganu. Journal of Sustainable Science and Management Special Issue Number 1: The International Seminar on the Straits of Malacca and the South China Sea, 2016, 11-18.

［7］Kier, G., Kreft, H., Lee, T. M., Jetz, W., Ibisch, P. L., Nowicki, C., Mutke, J. & Barthlott, W. (2009). A Global Assessment of Endemism and Species Richness Across Island and Mainland Regions. Proceedings of the National Academy of Sciences, 106 (23), 9322-9327.

［9］Peh, K. S. H., De Jong, J., Sodhi, N. S., Lim, S. L. H. & Yap. C. A. M. (2005). Lowland Rainforest Avifauna and Human Disturbance: Persistence of Primary Forest Birds in Selectively Logged Forests and Mixed-Rural Habitats of Southern Peninsular Malaysia. Journal of Biological Conservation, 123 (4), 489-505.

［14］Rodrigues, P. & Cunha, R. T. (2012). Birds as a Tool for Island Habitat Conservation and Management. American Journal of Environmental Science, 8 (1), 5-10.

［18］Sodhi, N. S., Koh, L. P., Prawiradilaga, D. M., Tinulele, I., Putra, D. D., Tan, T. H. T. (2005). Land Use and Conservation Value for Forest Birds in Central Sulawesi (Indonesia). Journal of Biological Conservation, 122 (4), 547-558.

［22］Taylor, S. & Kumar, L. (2016). Global Climate Change Impacts on Pacific Islands Terrestrial Biodiversity: A Review. Tropical Conservation Science, 9 (1), 203-223.

［23］Turner, C. S., King, T., O'Malley, R., Cummings, M. & Raines, P.(2002). Danjungan Islands Biodiversity Survey: Terrestrial Final Report. London, Coral Cay Conservation, 3-77.

［26］Yap, C. A. M., Sodhi, N. S. & Peh, K. S. H.(2007). Phenology of Tropical Birds in Peninsular Malaysia: Effects of Selective Logging and Food Resources. The Auk, 124 (3), 945-961.

练习题

1. 直接引用时，MLA 和 APA 的规则是怎样的？

2. 如何才能避免学术不端？

3. 阅读下面一段话，先说出它的引用格式，并复习该格式的相关内容。

Nowadays, due to serious water shortages, especially in arid and semiarid areas, the view toward waste water has changed and is now considered to be a valuable water resource. The growth of population in urban areas as well as increased agricultural and industrial activities implies the need for the utilization and preservation of the environment and human resources as well as waste water to meet the human needs (Ferguson 1999; Barragan et al. 2017). One of the major concerns of using waste water is its possible heavy metal contamination. Heavy metals, including Cd, Cr, Cu, Pb, Hg, are not biodegradable, so they can accumulate in the body of living organisms, causing cardiovascular diseases and abnormalities (Wei and Yang 2010). Therefore, some cost effective and environmentally friendly technology for the treatment of heavy metal-contaminated waste water is necessary. For example, in Italy, 1570 sites out of 9000 (Krpata et al. 2008) and, in Norway, 714 sites out of 3359 (Glass 1999) can be treated effectively.

Using municipal and industrial waste water for irrigating suburban areas has become common in many parts all around the world (Singh and Agrawal 2008). All kinds of waste water, including refined and unrefined, rainfall and run of from industrial and domestic waste water, may be needed to be used for irrigating urban parks and forest margins of the cities and industrial complexes (Anderson 1996; Shereif et al. 1995; Al-Jamal et al. 2000). In this case, tree species are able to absorb heavy metals through their developed root system and reduce soil toxicity. Therefore, they can play an important role in protecting the environment (Karpiscak et al. 1996; Stewart et al. 1990).

（本部分摘自 Maryam Safari Aman, Mohammad Jafari, Majid Karimpour Reihan and Babak Motesharezadeh 的文章 Assessing Some Shrub Species for Phytoremediation of Soils Contaminated with Lead and Zinc）

参考文献

[1] 黑玉琴，郭粉绒. 英语学术论文写作（本科用）[M]. 西安：西安交通大学出版社，2017.

[2] 美国现代语言协会. MLA 科研论文写作规范 [M]. 7 版. 上海：上海外语教育出版社，2011.

[3] 美国心理协会. APA 格式（社会科学学术写作规范手册英语第 6 版）/万卷方法 [M]. 席仲恩，译. 重庆：重庆大学出版社，2011.

[4] 全国文献工作标准化技术委员会第六分委员会. 文后信息与文献 参考文献著录规则：GB/T 7714—2005 [S]. 北京：国家标准局，2005.

[5] 中华人民共和国国家质量监督检验检疫总局，中国国家标准化管理委员会. 信息与文献 参考文献著录规则：GB/T 7714—2015 [S]. 北京：中国标准出版社，2015.

[6] Aman, Maryam S, et al. Assessing Some Shrub Species for Phytoremediation of Soils Contaminated with Lead and Zinc [J]. Environmental Earth Sciences, 2018, 77 (3): 82~91.

[7] Bulant and Klims. 3-D Velocity Models - Transformation from General to Natural Splines [J]. Studia Geophysica et Geodaetica, 2019, 63: 137~146.

[8] Gilbert, et al. An update on the bird population in Gaya island [J]. Transactions on Science and Technology, 2018, 5 (2): 171~176.

[9] 李慧颖. The Impact and the Solutions of English Linguistic Imperialism [J]. 海外英语, 2019 (3): 231~232.

[10] Lindquist, et al. The effects of dislocations on crystallographic twins and domain wall motion in magnetite at the verwey transition [J]. Earth, Planets and Space, 2019, 71: 5.

[11] Posner R A. The Little Book of Plagiarism [M]. New York: Pantheon, 2007.
[12] American Psychological Association. Publication manual of the American Psychological Association [M]. 6th ed. Washington DC: American Psychological Association, 2010.
[13] Tan, et al. Reclamation and reuse of abandoned quarry: a case study of ice world & water park in Changsha [J]. Tunneling and Underground Space Technology, 2019, 85: 259~267.
[14] Yao, et al. Effects of shaft inclination angle on the capacity of smoke exhaust under tunnel fire [J]. Indoor and Built Environment, 2019, 28: 77~87.
[15] Werker J, Wulf C, Zapp P. Working conditions in hydrogen production: a social life cycle assessment [J]. Journal of Industrial Ecology, 2019, 1: 10.

学术论文写作中的问题汇总

本章将从写作态度、论文结构与格式、论文内容、语言与文风四个方面对学术论文写作的常见问题加以汇总、分析,并给出写作建议。

7.1 学术不端问题

学术研究本质上是求真、求实和创新。违背这一本质的学术探索就容易走入歧途,导致各种学术不规范甚至学术不端行为。

7.1.1 学术不端行为界定

目前,我国对科研与学术出版过程中的道德伦理问题非常重视,对学术不端行为零容忍。相关部门针对包括学术研究、学术成果、学术传承、学术管理在内的整个学术体系的学术不端行为出台了一系列规章制度,旨在逐步完善国家和个人科研诚信体系。尤其是 2019 年 5 月 2 日国家新闻出版署发布的《学术出版规范——期刊学术不端行为界定》(CY/T 174—2019),对学术期刊论文作者、审稿专家、编辑者所可能涉及的学术不端行为进行了详细规范地界定,对于学术不端的预防和处理有着指导性的现实意义。

根据《界定》,涉及论文作者的学术不端行为类型主要包括[1]:

(1) 剽窃(Plagiarism):采用不当手段,窃取他人的观点、数据、图像、研究方法、文字表述等并以自己名义发表的行为。

(2) 伪造(Fabrication):编造或虚构数据、事实的行为。

(3) 篡改(Falsification):故意修改数据和事实使其失去真实性的行为。

(4) 不当署名(Inappropriate Authorship):与对论文实际贡献不符的署名或作者排序行为。

(5) 一稿多投(Duplicate Submission;Multiple Submissions):将同一篇论文或只有微小差别的多篇论文投给两个及以上期刊,或者在约定期限内再转投其他期刊的行为。

(6) 重复发表(Overlapping Publications):在未说明的情况下重复发表自己(或自己作为作者之一)已经发表文献中内容的行为。

(7) 其他学术不端行为:由"第三方"代写、代投论文;引用文献不规范不适当等。

---

[1] 同方知网数字出版技术股份有限公司,中国科学院科技战略咨询研究院. 学术出版规范——期刊学术不端行为界定(CY/T 174—2019)[S]. 北京:国家新闻出版署,2019.

7.1.2 原因分析与写作建议

学术不端问题有些性质恶劣，论文作者为了个人利益故意违规甚至违法，比如抄袭、剽窃他人劳动成果；有些是因为写作态度不认真敷衍了事，比如随意选择数据和引用资料、引用文献不规范；有些则是不了解学术不端具体规定，无意中违规，比如作者不知道在论文中重新使用自己之前发表过的图表而未注明出处属于自我剽窃行为。

无论故意还是无意，出现学术不端的根源在于：学术态度不端正不严谨，急功近利、心浮气躁、投机取巧、缺乏对学术的敬畏之心。

避免学术不端要做到：

（1）加强自身学术规范、学术道德教育与反思；端正学术态度，实事求是、踏踏实实做科研；遵纪守法，力奉学术诚信至上，尊重别人的知识产权与劳动成果。

（2）认真研读《学术出版规范——期刊学术不端行为界定》（CY/T 174—2019）。《界定》中对可能出现的每种不端行为的具体表现形式都做了详细解释，比如剽窃就包括观点剽窃、数据剽窃、图片和音视频剽窃、研究（实验）方法剽窃、文字表述剽窃、整体剽窃、他人未发表成果剽窃七种类型，每种类型又细分若干小项。每一小项都是论文写作过程中要避开的"雷"。

（3）认真研读我国关于参考文献的引用标准《信息与文献 参考文献著录规则》（GB/T 7714—2015）。熟悉引用规则，避免出处标注不规范。

（4）参考本领域权威杂志期刊上的文章，确定可以使用的正确引用格式，包括在文中的标识方式，以及文末参考文献的写作格式。

（5）用规范严谨的学术行为防止无意犯错。比如前期搜集信息资料阶段做好笔记摘抄，认真仔细完整地记录信息来源，避免出处标注疏漏、不准确或不完整。

（6）学会合理引用他人研究成果与文献资料。合理的引用可以把前人的研究成果作为支撑自己观点的有力论据纳入文中，不能为了避免剽窃嫌疑而对引用避之若虎。合理引用包括直接引用和间接引用，详细技巧与方法请参见第6章。

7.2 论文结构与格式问题

学术论文是科学界的"八股文"，在一个相对完整的结构内，按照一个基本固定的格式，借用大小标题和各种符号，把整个学术研究过程分成几个板块，按顺序对科学内容进行记录。

7.2.1 常见问题

结构方面的常见问题包括：

（1）论文结构混乱或者不完整：缺少必需板块；板块内缺少必要的内容；板块顺序前后颠倒；同时出现多种结构方式。

（2）结构松散：各板块缺少条理；子标题缺少逻辑关联；逻辑顺序词缺失或者使用不

准确。

（3）子标题与符号系列使用混乱。

（4）行文格式混乱、前后不一致，甚至出现格式错误：字体、字号、行间距、段落格式等。

7.2.2 原因分析与写作建议

出现结构问题主要因为不清楚论文的基本结构组成，不了解每个板块需要包含的内容，没有弄明白各板块的具体表达格式；再就是语法功底弱，文章架构能力有待加强。建议在写作前多读权威杂志论文，熟悉学术论文的结构与板块，每个板块包含的必需内容，以及图表、数据、引用等具体的写作格式。

（1）把握学术论文的基本内容结构和每个板块的具体表达内容。英语文章（Essay）基本包括三部分：引言（Introduction）→正文（Main Body）→结论（Conclusion）。引言部分负责提出问题，在正文部分分析问题，结论部分呼应引言，解决问题。学术论文如期刊文章（Article）其实还是遵循提出问题—分析问题—解决问题的逻辑思维过程，结构上也是符合引言—正文—结论的基本框架，只是被赋予了更具体化的格式，内容更复杂的正文部分被细分成几个板块。各板块也有自己的基本结构与写作顺序以表达出相对固定的内容。如果写毕业论文（Dissertation），还要加上序言（Preface）、目录（Contents）、致谢（Acknowledgement）和索引（Index）等板块。一般论文的结构见表7-1。

表7-1 论文的基本结构与内容

标题（Title）	内容
摘要（Abstract）	研究背景与研究问题 → 研究目的与重要性 → 研究方法 → 研究结果→结论
引言（Introduction）	研究背景 → 研究问题 → 研究现状 → 研究目的
文献综述（literary review）	
正文（Main Body）	材料和方法（Materials & Methods）→ 研究结果（Findings）→讨论（Discussion）
结论（Conclusion）	本研究的结果及说明的问题→本研究对以前观点的贡献（创新点）→本研究的不足之处及后续研究方向
参考文献（Bibliography/References）	
附录（Appendices）	

（2）按照论文基本结构列出提纲，写出每一项的内容关键词。

（3）确定论文结构层次。如果用自然段，就写出每个板块的主题句；如果用小标题就需要按照格式要求进行分层，拟就大小标题，并把大小标题按层次进行连续编码。以本章部分标题为例，见表7-2。

表 7-2 论文格式要求示例

7　学术论文写作中的问题汇总	一级标题
7.1　学术不端问题	二级标题
7.1.1　学术不端行为界定	三级标题
7.1.2　原因分析与写作建议	三级标题
7.2　论文结构与格式问题	二级标题
7.2.1　常见问题	三级标题
7.2.2　原因分析与写作建议	三级标题
7.3　论文内容方面的问题	二级标题
7.3.1　常见问题	三级标题
7.3.2　原因分析	三级标题
7.3.3　写作建议	三级标题
7.4　论文语言与文风问题	二级标题
7.4.1　常见问题	三级标题
7.4.1.1　文风问题	四级标题
7.4.1.2　语言问题	四级标题
7.4.2　原因分析与写作建议	三级标题

（4）句法上运用表示逻辑顺序的词来承上启下，实现句子与句子，段落与段落，板块与板块的环环相扣，使文章有机结合成一个整体。尤其在列举材料和方法、研究结果、问题讨论时，运用表示顺序、因果、转折等关系的标志语可以使整个实验过程的流程记录严谨有序。

（5）认真研读《科学技术报告/学位论文和学术论文的编写格式》，并按照出版者或命题者给出的格式要求，在格式上统一全文。特别注意图表、注释和参考文献部分的要求格式，严谨细致地标注。

7.3　论文内容方面的问题

一篇结构上完完整整，方方面面都列出来的学术论文依然不能保证是一篇好论文。内容充分而不冗余，论证充足而准确，观点独到而创新，这才是学术论文的灵魂。一篇好的学术论文内容上会兼具思想性，创新性和科学性。读者能从中了解到某领域最新研究现状与不足，论文提出问题、解决问题的方法过程，论文创新点等。论文每个板块的具体写作方法前面章节有详述，这里不再赘述。

7.3.1　常见问题

（1）标题。

1）标题宽泛不具体。

2）表意不明确，不能概括论文的主要观点。

3）标题冗长，含无信息词汇，无重点。

4）书写不规范。

（2）摘要。

1) 摘要与论文内容不紧密。
2) 只是正文中子标题的简单罗列。
3) 照搬引言或结论。
4) 出现前言、背景、文献综述、例证、研究过程等不必要的内容。
5) 出现评论与注释。
6) 出现文学修饰。
7) 出现特殊字符。
8) 出现图表和化学结构式数学表达式。
(3) 关键词。
1) 术语过长。
2) 词目过多。
3) 不具有典型性，没有高度概括论文。
(4) 正文部分。
1) 内容不科学。
2) 缺少充分论据、内容空洞无物。
3) 东拼西凑、充斥大量无用信息。
4) 流水账式罗列事实数据，重点不突出。
5) 论据不足或与论点无关。
6) 分析与论证缺乏说服力。
7) 没有创新点。
(5) 图表。
1) 图表使用不当：多用、乱用或者缺少。
2) 图表质量不高。
3) 图文不符：与论文内容不符或与文字描述不符。
4) 图表制作不规范，设计不合理。
(6) 结论。
1) 与文章主要内容没有紧密关联，没有呼应引言部分提出的问题。
2) 生硬地提出或拔高创新点，没有说服力。
3) 对论文成果没有进行具体客观地分析评价。

7.3.2 原因分析

(1) 前期准备工作不足。巧妇难为无米之炊。论文写作是个输出过程，写作之前需要有一个文献收集、阅读、研究分析的输入过程，输入的缺失或者输入量不足必然导致无内容可写。具体包括：1) 没有收集足够多足够权威的文献资料；2) 对资料的整理与研究不充分，对本领域之前的研究工作没有整体把握，没有具体数据支撑说明同类工作的既有成果；3) 调研取材不具有普遍性和代表性；4) 缺少实验数据支撑，没有第一手研究结果，没有自己的实验数据。

(2) 有关学术论文写作的理论修养不足。具体包括：1) 不会筛选材料，引用的资料没有说服力；2) 只罗列现象，缺少深层挖掘，研究结果解释的不充分，显示不出论文中自己的研究成果的核心价值与重要意义。

7.3.3 写作建议

（1）论文写作前做好充分的准备工作。美国国家基金会在化工部的调查统计表明，科研人员工作时间分配为：计划思考 7.7%，信息搜集 50.9%，实验研究 32.1%，数据处理 9.3%。前期准备工作做得充分，论文写作是水到渠成的事情。写论文的一般流程为：

1）信息搜集：利用图书馆、专著、学术期刊、学术会议论文集、数据库、网络资源和其他相关资源，围绕计划写作的主题，阅读大量相关文献，特别是近期发表的权威刊物高水平学术文章。了解行业内技术发展的状况，收集论文相关资料。

2）资料分析：对收集到的资料进行评估筛选、做好摘抄、分类整理。

3）实验研究：记录一手实验原始数据、实验过程。

4）实验数据处理：结果与分析。

5）借鉴本领域权威刊物论文格式，确定论文将要采用的格式。

6）编写提纲：固定论文采用的结构和要描述的内容要点。

7）按照论文的选定结构开始逐个板块地进行写作。

8）修订校对：初稿完成，需要再从头开始看看是否说清楚了你研究的问题，确保没有跑题。

（2）标题要反映论文的中心内容和主要观点，同时提供检索所需的特定实用信息。标题需要在论文进行前就拟定，但论文成文后修订过程中一定要根据全文内容，特别是根据摘要和关键词对其做最后调整。最终确定的标题要简洁、准确、醒目且信息量大，包括与主题相关的最重要的词汇。

（3）摘要出现在论文最前面，但它其实是在文章全文完成并提炼出关键词后再通览全文拟就的。它是对文章主要内容与写作目的的高度概括。它用非常简练的语言，提炼出论文的研究背景、研究目的、研究问题、研究方法、研究结果和论文创新点。整个摘要的长度一般不能超过文章内容的 5%，大约 200~250 字。

（4）关键词是代表全文主要内容的单词或术语，为文献标引检索服务。从论文标题或者正文中选取 3~8 个与文章论点紧密相关的词作为关键词。

（5）正文的引言部分要说明研究背景、研究问题、研究现状、一个含义明确的研究性问题以及你将如何在论文中回答这个问题。材料和方法、研究结果、问题讨论等内容是遵循严谨的结构层次对整个实验论证过程的客观、精准描述。这几部分一定是依托于严密合理的科研设计、正确的研究方法、完整可靠的资料、充分准确的数据等，通过充分的证据，得出有说服力的结论，在一定程度上获得创新。正文部分要用较长的篇幅展开论证分析，如果作者调研不足或逻辑思维不强，最受影响的板块就是正文部分。一定要熟悉各种逻辑顺序词的使用并加以练习。

（6）图表作为论文重要的论据，必须保证数据要准确，格式正确，与论点紧密相关。要有图表的描述内容与数据分析内容。

（7）结论部分不是对引言的简单重复，而是对全文论点的重申、呼应以及拔高。这部分内容主要包括：1）总结本研究的结果及说明的问题；2）评述本研究对以前观点的贡献（创新点）；3）本研究的不足之处及后续研究方向。

7.4 论文语言与文风问题

学术论文是应用文,要遵循清晰简洁的原则,以功能性为本,语言表达的基本要求是正式、客观、精确、简洁、切题,重点突出层次分明;文体符合一定的格式规范。语言表述不清,用词不当等语言问题必然影响学术成果的传播。

7.4.1 常见问题

7.4.1.1 文风问题

(1) 主观评论较多。出现表明个人视角个人态度的主观词汇和一些绝对语气词。
(2) 语言啰嗦,机械重复,空话套话,多余的交代与说明。
(3) 事无巨细,流水账式记录。
(4) 语言枯燥,词汇贫乏。
(5) 语义晦涩,生硬难懂,不知所云。
(6) 华丽辞藻堆砌。

7.4.1.2 语言问题

(1) 人称、时态、语态问题:论文中随意切换第一人称和第三人称,写作风格不一致;时态使用不恰当。
(2) 句式繁琐,冗长花哨。
(3) 句子有歧义。
(4) 句子支离破碎,语法规则使用错误。
(5) 图表问题:主要表现在格式不清楚,不会对图表进行描述、不会进行数据分析。
(6) 语言表达不正式、不书面、不精确、不简洁、不客观、带有主观色彩。
(7) 词汇使用不当:词汇意义不准确,专业术语使用不当,搭配不当,杜撰新词,中式英语,缩略词使用不当,滥用大词,冠词使用不当。
(8) 标点符号运用不当。

7.4.2 原因分析与写作建议

语言基本功不够,学术论文写作功力不足,不了解论文的本质,不熟悉论文写作格式。整体上提高自己的学术论文写作水平,需要注意三点:

(1) 加强自己的语言基本功训练。
(2) 熟悉国家学术论文写作格式标准。
(3) 参考本领域权威期刊上母语作者的论文。观察他们用什么句式来表达,怎样使用图表,并加以模仿。可以考虑使用一些简单的规则和模板。

虽然每一篇学术论文的具体风格不能一概而论,但是依然有一些一般规则可以遵循:

(1) 语言风格朴实、不浮夸。不追求花哨的语言繁琐的句式,少用表现极端感情的词汇。
(2) 人称、语态、时态。人称前后尽量一致;时态一般采用过去时和现在时,不用将来时;不滥用被动语态。

以前的学术论文倾向于使用第三人称被动语态，现在则主张使用主动的第一人称进行学术写作，简洁有力，动作发出者明确，能更有效地传达信息和吸引读者。当我们无法明确动作的发出者，或者需要强调动作本身而不是强调动作发出者的时候，才去使用被动语态。尤其是摘要部分，少用被动语态和第一人称，用"the study""this paper examines"or"this research"来代替"my study"or"I write about……"。

（3）长短句搭配使用。短句简洁准确、清楚易懂。单纯短句的堆砌会导致文章枯燥；长句可以缩短文章，但是长句不容易架构，刻意追求长句又会显得文章冗长花哨，建议长短句搭配使用，多用短句。S. Bailey 在他的著作中列举了一些很实用的简练句子小窍门[1]：

1）用动词代替名词。英语中名词往往需要和动词连用，不如直接用动词简洁。比如：要表达"做出贡献"，直接用 contribute 代替 make contributions to。

2）尽量避免使用动词词组。比如：
把 bring up, go on 改成 raise, continue。

3）不用 get 词组。比如：
get better/worse 改成 improve, deteriorate。

4）少用 there be 句型。比如：
There is a necessity for a semi-structured approach to be chosen. 改成：A semi-structured approach must be chosen.

5）精简形式主语 it。比如：
It is essential that the model be revised. 改成：The model must be revised.

6）如果句子比较长，可以在其中加两三个逗号分隔，太长的句子，则建议拆成两三个短句。

（4）以陈述句为主，不滥用感叹句和疑问句。比如：
把 What were the reasons for the deadline in wool exports? 改成：There were four main reasons for the deadline.

（5）注意语法与句法，避免句子产生歧义。

1）比较句型形式要保持完整，省略被比较主体后面的动词，有时会产生歧义。比如：
把 Country A funds high-tech innovation more than Country B. 改成：Country A funds high-tech innovation more than Country B does.

2）修饰语位置要放在被修饰的词后面，错置容易有歧义。比如：
I was told that I would be awarded the scholarship by my professor. 修饰语 by my professor 这一修饰语放置不当导致这句话两种理解：我被教授告知我会获得奖学金，我被告知我的教授会授予我奖学金。

3）代名词使用要谨慎，必要时可对代名词进行替换。比如：
A case study approach was chosen; this allowed a closer observation of a single specimen. 句子中 this 可能指代 a case study approach，也可能指代 a case study approach was chosen 这一整句话，为了使句意清晰，我们可以将句子改为：A case study approach was chosen to allow a closer observation of a single specimen.

---

[1] Stephen bailey. Academic Writing [M]. Second version. New York：Routledge, 2006：106~107.

（6）使用正式、标准的语言，避免口语化、习语类表达。比如：

1）用 small、large、man、mother 代替 little、big、guy、mom 这类口语化的词。

2）动词的否定形式不用省略式，用 do not、can not，不用 don't、can't。

3）引入例子用 such as、for instance，不用 like。

4）用 positive、negative 代替 good、bad 这类非常简单的词。

（7）遣词造句选择准确、清楚的语言，避免不精确与指代模糊。比如：

1）词汇 rule 和 law 意义是不一样的。

2）用具体的 reduce 或者 increase 来代替 influence。

3）用 factor、issue、topic 代替 thing、nothing、something。

4）用 significant/considerable number 代替 lots of。

5）列举时避免使用 etc. And so on，在最后两项中间加上 and。

（8）遣词造句选择客观描述的语言，避免带感情色彩的词汇和绝对语气词。比如：

1）避免使用 I think，I believe，according to my experience 等主观描述的表达方式。

2）避免使用 Luckily，remarkably，surprisingly 等类型的绝对语气词。

3）把 Technology raises the risk of emotional laziness. 改成：Technology may raise the risk of emotional laziness. 或者 Technology tend to raise the risk of emotional laziness.

（9）缩略词用法。可以使用众所周知的缩略词，比如 WTO（World Trade Organization）。可以使用学术论文中常见的方便易懂的缩略词，比如：et al. = and others，Fig. = figure，i. e. = that is，cf. = compare，e. g. = for example。避免使用存在一词多义的缩略词，比如 AIDS 这个缩写就对应着自动识别系统、人工免疫系统和会计信息系统三个含义。特定研究领域的缩略词，第一次使用的时候需要给出全称来做解释。

（10）避免使用可能会造成性别歧视的代名词 he，she，his，her，him 等。比如：

把 When a politician campaigns for office, he must spend considerable funds to compete with his opponents. 改成：When politicians campaign for office, they must spend considerable funds to compete with their opponents. 或者：A politician who campaigns for office must spend considerable funds to compete with opponents.

（11）图表描述与分析。Figure 和 Table（diagram，table，map，pie chart，bar chart，line graph，etc.）都应该有数字排序和标题。Table 的标题在上面，Figure 的在下面。图表描述与分析有约定俗成的句式，可以套用模仿，具体句式可参考前面章节。

最后，学术论文写作是一个积累的过程，需要有耐心，多读文献、多读论文、加以模仿参考，从内容、结构、语言三方面多加磋磨练习。英语有这样一句谚语"模仿是最真诚的赞美"，这句话用在学术论文写作上也完全适用。

练习题

1. S. Bailey 在其著作 Academic Writing 中用下面两个段落来介绍学术论文写作要求[1]。他认为和段落 A 比较，B 更符合学术论文写作规范。请对比两个段落的不同之处，指出 A 段存在的问题。

段落 A：A lot of people think that the weather is getting worse. They say that this has been going on for quite a

---

[1] Stephen bailey. Academic Writing [M]. Second version. New York：Routledge，2006：105.

long time. I think they are right. Research has shown that we now get storms etc all the time.

段落 B：It is widely believed that the climate is deteriorating. It is claimed that this process has been continuing for nearly 100 years. This belief appears to be supported by Mckinley (1997) who shows a 55% increase in the frequency of severe winter gales since 1905.

2. 下面这段摘要选自 Science 期刊的论文 Polyamide membranes with nanoscale Turing structures for water purification❶。此摘要一共包括五句话，这五句话分别对应的是摘要的哪部分内容？

The emergence of Turing structures is of fundamental importance, and designing these structures and developing their applications have practical effects in chemistry and biology. We use a facile route based on interfacial polymerization to generate Turing-type polyamide membranes for water purification. Manipulation of shapes by control of reaction conditions enabled the creation of membranes with bubble or tube structures. These membranes exhibit excellent water-salt separation performance that surpasses the upper-bound line of traditional desalination membranes. Furthermore, we show the existence of high water permeability sites in the Turing structures, where water transport through the membranes is enhanced.

3. 下面这个段落和参考文献部分选自 Science 期刊的论文 Polyamide membranes with nanoscale Turing structures for water purification❶。请分析：1) 这个段落应该是论文的哪一板块；2) Science 期刊的引用格式；3) 这个段落中出现了哪种或哪几种引用方式。

Alan Turing's 1952 paper (1),"The chemical basis of morphogenesis," theoretically analyzed how two chemical substances, activator and inhibitor (2) (Fig. 1A), can, under certain conditions, react and diffuse with each other to generate spatiotemporal stationary structures. Turing's ideas have profoundly influenced theoretical understanding of pattern formation in chemical (3) and biological (4, 5) systems, but it was not until nearly 40 years after his paper was published that experimental evidence was obtained for the chlorite-iodide-malonic acid (CIMA) reaction (6, 7). About 10 years later, stationary Turing states were also observed in the Belousov-Zhabotinsky (BZ) reaction microemulsion consisting of reverse micelles (8). Recently, a variety of two- and three-dimensional stationary structures were studied in chemical (9, 10) and biological (11~15) systems.

REFERENCES AND NOTES

1. A. M. Turing, Philos. Trans. R. Soc. Lond. B Biol. Sci. 237, 37-72 (1952).

2. A. Gierer, H. Meinhardt, Kybernetik 12, 30-39 (1972).

3. G. Nicolis, I. Prigogine, Self-organization in Nonequilibrium Systems (Wiley, 1977).

4. H. Meinhardt, Models of Biological Pattern Formation (Academic Press, 1982).

5. J. D. Murray, Mathematical Biology (Springer, 1989).

6. V. Castets, E. Dulos, J. Boissonade, P. De Kepper, Phys. Rev. Lett., 64, 2953-2956 (1990).

7. Q. Ouyang, H. L. Swinney, Nature 352, 610-612 (1991).

8. V. K. Vanag, I. R. Epstein, Phys. Rev. Lett. 87, 228301 (2001).

9. J. Horváth, I. Szalai, P. De Kepper, Science 324, 772-775 (2009).

10. T. Bánsági Jr., V. K. Vanag, I. R. Epstein, Science 331, 1309-1312 (2011).

---

❶ Zhe Tan, et al. Polyamide membranes with nanoscale Turing structures for water purification [J]. Science, 2018 (360): 518~521.

11. S. Sick, S. Reinker, J. Timmer, T. Schlake, Science 314, 1447–1450 (2006).
12. S. Kondo, T. Miura, Science 329, 1616–1620 (2010).
13. P. Müller et al., Science 336, 721–724 (2012).
14. R. Sheth et al., Science 338, 1476–1480 (2012).
15. J. Raspopovic, L. Marcon, L. Russo, J. Sharpe, Science 345, 566–570 (2014).

参考文献

[1] 同方知网数字出版技术股份有限公司，中国科学院科技战略咨询研究院. 学术出版规范——期刊学术不端行为界定（CY/T 174—2019）[S]. 北京：国家新闻出版署，2019.

[2] Stephen bailey. Academic Writing [M]. Second version. New York：Routledge，2006：105~107.

[3] 余丽，陈洁，梁永刚. 学术英语写作 [M]. 北京：清华大学出版社，2019.

[4] 马莉. 英语学术论文写作及语体风格 [M]. 北京：北京大学出版社，2017.

[5] 史宝辉，李芝. 英语学术论文写作教程 [M]. 北京：中国人民大学出版社，2018.

[6] 刘振聪，修月祯. 英语学术论文写作 [M]. 2版. 北京：中国人民大学出版社，2013.

[7] Tan Zhe, Chen Shengfu, et al. Polyamide membranes with nanoscale Turing structures for water purification [J]. Science，2018（360）：518~521.

附　　录

中文科技论文参考文献的运用及其著录规则

一、参考文献定义及其作用

按照《信息与文献　参考文献著录规则》(GB/T 7714—2015)的定义,参考文献(Reference)是对一个信息资源或其中一部分进行准确和详细著录的数据,位于文末或文中的信息源。因此参考文献的引用既是科学研究的起点和基础,也是论著的重要组成部分,是作者对他人研究成果的理论、观点、资料和方法的引用和借鉴,对文章内容起着支持、佐证和解释信息来源的作用,具有提供依据、分享研究成果、重复利用和可检索等重要功能。关于著录参考文献的意义及其作用,已有众多文献做了论述[1-6],现归纳如下:

(1) 反映研究者的研究基础,体现科学的传承性。
(2) 尊重前人和他人的知识成果,有利于知识产权的保护。
(3) 利于相关文献的进一步检索,实现信息资源共享。
(4) 精练文字,节省论文篇幅。
(5) 有利于通过引文分析对期刊水平做出客观评价。
(6) 促进科学情报和文献计量学研究,推动学科发展。

二、参考文献引用的原则[7-8]

(1) 直接引用的文献。要求著录的是论文作者亲自阅读过并在论文中直接引用的文献,不能间接引用他人文献后引用的参考文献,即尽量不要引用二次文献。

(2) 引用最新、最必要的文献。引用的参考文献越新,就越能反映该领域研究的最新动向和成果,从而使其科研选题更具有前沿性;研究者在众多的文献中要精选,要选择最主要的、最有效的文献,以突出重点,所引用的参考文献还应有充分的数量以保证信息来源的全面性。

(3) 引用正式出版的文献。研究者所引用的文献必须是国内外公开出版发行的文献,未公开发表的资料不应作为参考文献。

(4) 采用标准化的著录格式。论文作者应熟练掌握《信息与文献　参考文献著录规则》(GB/T 7714—2015),并按此规则或出版社、编辑部等要求的格式进行著录。

三、参考文献的分类

参考文献按照不同的标准可以划分为不同的类型。新修订的《信息与文献　参考文献著录规则》(GB/T 7714—2015)对"参考文献"做了明确的定义并将参考文献区分为"阅读型参考文献(Reading Reference)"和"引文参考文献(Cited Reference)",并对其标注和著录做了具体规定。阅读型参考文献是指"著者为撰写或编辑论著而阅读过的信息资源，或供读者进一步阅读的信息资源"；引文参考文献是指"著者为撰写或编辑论著而引用的信息资源"[9]。

四、参考文献类型和标识代码

文献类型标识是标示各种参考文献类型的符号。论文著者应严格按照《信息与文献　参考文献著录规则》（GB/T 7714—2015）的规定，将自己引用的各种参考文献的类型及载体类型用对应的文献类型标识码标示出来，见附表1和附表2。

附表1　文献类型及其标识代码

参考文献类型	文献类型标识代码
普通图书	M
会议录	C
汇编	G
报纸	N
期刊	J
学位论文	D
报告	R
标准	S
专利	P
数据库	DB
计算机程序	CP
电子公告	EB
档案	A
舆图	CM
数据集	DS
其他	Z

附表2　电子资源载体和标识代码

电子资源的载体类型	载体类型标识代码
磁带（Magnetic Tape）	MT
磁盘（Disk）	DK
光盘（CD-ROM）	CD
联机网络（Online）	OL

五、参考文献著录项目及著录要求[9]9-12

（一）主要责任者（Creator）或其他责任者

主要责任者可以是个人或团体，包括著者、编者、学位论文撰写者、专利申请者或专利权人、报告撰写者、标准提出者、析出文献的著者等。

（1）个人著者采用姓在前名在后的著录形式。用汉语拼音书写的人名，姓全大写，其名可以缩写，取每个汉字拼音的首字母[10]；欧美著者的名可用缩写字母，缩写名后省略缩写点，名字间间隔一个字符。国际主流科技期刊中欧美著者的姓名顺序通常是名前姓后，如"Vicki Wilmer"，根据新标准，正确著录格式应为"WILMER V"。"Albert Einstein"应著录为"EINSTEIN A"。有的著者是复姓，如"Amabel Williams-Ellis"，不能著录为"Williams Amabel"或者"Ellis Amabel"，正确的著录格式为"WILLIAMS-ELLIS A"。欧美著者中译名"只著录姓"，同姓不同名者"既著录姓，又著录名"。

（2）同一文献的责任者不超过3人时，全部著录；超过3人时，只著录前3名作者，中文文献其后加"，等"，英文文献其后加"，et al"。不同作者姓名之间用"，"隔开，不用"和""and"等连词。

示例如下：

马克思，恩格斯

安秀芬，黄晓鹂，张霞

曹秀堂，陈波，郭建刚，等

YELLAND R L, JONES S C, EASTON K S, et al.

（3）无责任者或责任者情况不明的文献，"主要责任者"项要注明"佚名"或相应的词。采用顺序编码制编著的参考文献，可省略此项，直接著录题名。

（4）对文献负责的机关团体名称，一般根据著录信息源著录。用拉丁文书写的机关团体名称，应由上至下分级著录。

（二）题名

题名包括书名、刊名、报纸名、专利题名、报告名、标准名、学位论文名、档案名、舆图名、析出的文献名等。题名按著录信息源所载的内容著录。

示例如下：

西方文明史：问题与源头

台山核电厂海水库护岸抗震分析与安全性评价研究报告

轨道火车及高速轨道火车紧急安全制动辅助装置

北京理工大学（社会科学版）

（1）同一责任者的多个合订题名，著录前3个合订题名。不同责任者的多个合订题名，只需著录第一个或处于显要位置的合订题名。参考文献中不著录并列题名。

（2）文献类型标识（含文献载体标识）依据《信息与文献 参考文献著录规则》（GB/T 7714—2015）著录。电子资源既要著录文献类型标识，也要著录文献载体标识。参见附表1和附表2。

（3）其他题名信息包括副题名，说明题名文字，多卷书的分卷书名、卷次、册次，专利号，报告号，标准号等，可根据信息资源外部特征的具体情况决定取舍。

（三）版本

第 1 版不著录，其他版本说明应著录。版本用阿拉伯数字、序数缩写形式或其他标识表示。

示例如下：
2 版
新 1 版
4th ed.
Rev. Ed.

（四）出版项

出版项应按出版地、出版者、出版年顺序著录。

出版地指出版者所在地的城市名称。对同名异地或人们不熟悉的城市名，应在城市名后附省、州名或国名等限定语。

示例如下：
北京：高等教育出版社，1982.
New York：Academic Press，2012.

（五）页码

专著或期刊中析出文献的页码或引文页码，应采用阿拉伯数字著录。阅读型参考文献的页码著录文章的起讫页或起始页，引文参考文献的页码著录引用信息所在页。

示例如下：
谭丙煜．怎样撰写科学论文［M］．沈阳：辽宁人民出版社，1982：63.
刘国钧，王连成．图书馆史研究［M］．北京：高等教育出版社，1979：15～18，31.

（六）获取和访问路径

根据电子资源在互联网中的实际情况，著录其获取和访问路径。

示例如下：
王明亮．关于中国学术期刊标准化数据库系统工程的进展［EB/OL］．http：//www.cajcd.edu.cn/pub/wml.Txt/980810 2.html.1998-08-16/1998-10-04.

六、参考文献标注法[9]13-14

参考文献的标注体系有"顺序编码制（Numeric References Method）"和"著者-出版年制（First Element and Date Method）"两种著录体系。顺序编码制是指引文采用序号标注，参考文献表按引文的序号排序；著者-出版年制是指引文采用著者-出版年标注，参考文献表按著者字顺和出版年排序。引文参考文献既可以集中著录在文后或者书末，也可以分散著录在页下端。阅读型参考文献著录在文后、书的各章节和书末。我国出版物普遍采用的是顺序编码制，本章主要介绍顺序编码制。

顺序编码制是按正文中引用的文献出现的先后顺序连续编码，将序号置于方括号中。

（一）引用单篇文献，序号置于方括号中

示例[11]：21世纪以来，我国的地铁建设已经进入了快速发展时期。盾构法因其地层适应性强、速度快、安全性高等优点[1]，逐步成为城市地铁工程建设中的主要形式。然而，隧道的盾构开挖[2]过程及后期固结沉降导致地表不同程度的沉降变形，引起其上及邻近建筑物开裂、倾斜或坍塌[3]。因此，合理地分析预测盾构施工对邻近建筑物沉降的影响，划定隧道明显影响区域，加强对区域内建筑物的监测和采取必要措施，对保护建筑物具有重要意义[4]。

参考文献：

[1] 刘少楠. 郑州地铁5号线盾构施工对地表民用建筑沉降影响研究［D］. 郑州：河南工业大学，2016.
[2] 崔耀. 盾构法与大断面矿山法隧道并行近接施工的影响分区研究［D］. 成都：西南交通大学，2018.
[3] 孙长军，张顶立，郭玉海，等. 大直径土压平衡盾构施工穿越建筑物沉降预测及控制技术研究［J］. 现代隧道技术，2015，52（1）：136～141.
[4] 祁建军. 上软下硬地层中隧道施工对建筑物的沉降影响及预防措施研究［J］. 铁道建筑技术，2017，34（9）：84～88.

（二）同一处引用多篇文献时，应将各篇文献的序号在方括号内全部列出，各序号间用"，"，如遇连续序号，起讫序号间用短横线连接

示例：参考文献是科技论文不可或缺的组成部分，按照规范要求正确著录参考文献是科技论文作者的重要责任。科技论文作者往往不重视参考文献的著录，认识不到其重要性，缺少严谨对待参考文献的态度，以及对著录标准不了解不学习，导致参考文献的著录从格式到内容的不规范、甚至错误现象比比皆是。很多编辑同仁对参考文献著录的标准进行了探讨[2-6]，对很多著录错误进行了总结[7-10]，对于推动正确著录参考文献起到重要促进作用。

参考文献：

[2] 黄城烟，王春艳. 参考文献新国标若干重要概念的理解和著录方法［J］. 编辑学报，2016，28（3）：239～242.
[3] 祝清松. 基于GB/T 7714—2015的参考文献著录趋势分析［J］. 编辑学报，2016，28（4）：352～353.
[4] 朱大明. 阅读型参考文献与引文参考文献的概念及特征［J］. 编辑学报，2016，28（4）：324～326.
　　……
[10] 刘东信. 英文参考文献审校中需要注意的问题例析［J］. 编辑学报，2011，23（4）：316～317.

（三）多次引用同一著者的同一文献时，在正文中标注首次引用的文献序号，并在序号的"［　］"外著录引文页码

示例[12]：……为方便比较，笔者在3类参考文献格式中分别选择了具有代表性的2种，包括芝加哥格式（第16版，仅比较其注释—参考文献表格式）[5]708-748、MHRA格式（第3版）[25]、APA格式（第6版）[6]192、MLA格式（第7版）[7]148-180、NLM格式（第2版）[23]123-288和IEEE格式[26]。

参考文献:

[5] 美国芝加哥大学出版社. 芝加哥手册:写作,编辑和出版指南:第16版 [M]. 吴波,等译. 北京:高等教育出版社, 2014.

[6] 美国心理协会. APA格式:国际社会科学学术写作规范手册 [M]. 席仲恩,译. 重庆:重庆出版社, 2011.

[7] 现代美国语言协会. MLA科研论文写作规范:英文 [M]. 上海:上海外语教育出版社, 2011.
……

[23] Patrias K. Citing medicine:the NLM style guide for authors, editors, and publishers [M/OL]. 2nd edition. Bethesda (MD):National Library of Medicine (US), 2015. [2019-05-13]. https://www.ncbi.nlm.nih.gov/books/NBK7256/pdf/Bookshelf_NBK7256.pdf.

七、著录范式

正确著录参考文献的根本是要掌握《信息与文献 参考文献著录规则》(GB/T 7714—2015) 的要求,即:参考文献的著录内容应准确,必备项目不应缺失,著录项目的书写、著录顺序、著录用符号、引文处标注等应规范。根据此规则,结合各出版社、编辑部等要求的格式,各类文献著录格式要求归纳如下。

(一) 普通图书

格式:[序号] 主要责任者. 题名:其他题名信息 [M]. 其他责任者. 版本项. 出版地:出版者, 出版年:引文页码.

示例:

[1] 葛家澍, 林志军. 现代西方财务会计理论 [M]. 厦门:厦门大学出版社, 2001:42.

[2] AGRAWAL G P. 非线性光纤光学 [M]. 胡国绛, 黄超, 译. 天津:天津大学出版社, 1992:179~193.

[3] 中国社会科学院语言研究所词典编辑室. 现代汉语词典 [M]. 修订本. 北京:商务印书馆, 1996:258~260.

(二) 论文集、会议录

格式:[序号] 主要责任者. 题名:其他题名信息 [C]. 出版地:出版者, 出版年.

示例:

[1] 伍蠡甫. 西方文论选 [C]. 上海:上海译文出版社, 1979.

[2] 雷光春. 综合湿地管理:综合湿地管理国际研讨会论文集 [C]. 北京:海洋出版社, 2012.

(三) 学位论文

格式:[序号] 主要责任者. 题名 [D]. 培养单位所在地:培养单位, 出版年:引文页码.

示例:

[1] 吴明喜. 人工合成透明砂土及其三轴试验研究 [D]. 大连:大连理工大学, 2006.

［2］张筑生. 微分半动力系统的不变集［D］. 北京：北京大学数学系数学研究所，1983：1~7.

（四）报告

格式：［序号］主要责任者. 题名：其他题名信息［R］. 出版地：出版者，出版年.

示例：

［1］孔宪京，邹德高，徐斌，等. 台山核电厂海水库护岸抗震分析与安全性评价研究报告［R］. 大连：大连理工大学工程抗震研究所，2009.

（五）专利文献

格式：［序号］专利申请者或所有者. 专利题名：专利号［P］. 公告日期或公开日期［引用日期］.

示例：

［1］周成军. 建筑施工抹灰装置：CN201720456416.3［P］. 2017-12-22.
［2］天九城市森林花园建筑科技成都有限公司. 城市森林花园建筑及建筑群：CN201510849818.5［P］. 2017-06-06.

（六）标准文献

格式：［序号］主要责任者. 标准名称：标准号［S］. 出版地：出版者，出版年：引文页码.

示例：

［1］中国建筑科学研究院. 建筑地基基础设计规范：GB 50007—2011［S］. 北京：中国建筑工业出版社，2012.
［2］中华人民共和国国家质量监督检验检疫总局，中国国家标准化管理委员会. 信息与文献　参考文献著录规则：GB/T 7714—2015［S］. 北京：中国标准出版社，2015.

（七）专著中的析出文献

格式：［序号］析出文献主要责任者. 析出文献题名［文献类型标识］//析出文献其他责任者或专著主要责任者. 专著题名：其他题名信息. 版本项. 出版地：出版者，出版年：析出文献的页码.

注：专著中析出文献时需在析出文献题名和专著主要责任者之间加用"//"，并列出版本项（第1版不著录）。

示例：

［1］程根伟. 1998年长江洪水的成因与减灾对策［M］//许厚泽，赵其国. 长江流域洪涝灾害与科技对策. 北京：科学出版社，1999：32~36.
［2］李梢. 中医证候与分子网络调节机制的可能关联［C］// 周光召. 面向21世纪的科技进步与社会经济发展：中国科学技术协会首届学术年会. 北京：中国科学技术出版社，1999：442.

（八）期刊中析出的文献

格式：［序号］主要责任者. 题名：其他题名信息［J］. 期刊名，出版年，卷号（期号）：引文页码.

示例：

［1］夏鲁惠. 高等学校毕业设计（论文）教学情况调研报告［J］. 高等理科教育，2004（1）：46~52.
［2］曹秀堂，陈波，郭建刚，等. 医院学科评估方法与应用［J］. 解放军医院管理杂志，2009，16（2）：110~112.

（九）报纸中析出的文献

格式：［序号］主要责任者. 题名：其他题名信息［N］. 报纸名，出版日期（版面数）.

示例：

［1］谢希德. 创造学习的新思路［N］. 人民日报，1998-12-25（10）.
［2］李四光. 中国地震的特点［N］. 人民日报，1988-08-02（4）.
［3］叶继元. 从学术规范的视角论"好的学者"与"好的研究"［N/OL］. 光明日报，（2018-01-02）［2018-02-26］. http：//news. gmw. cn/2018-01/02/content_ 27242649. htm.

（十）电子文献著录范式

注：此电子文献不包括电子专著、电子连续出版物、电子学位论文、电子专利。

格式：［序号］主要责任者. 题名：其他题名信息［文献类型标识/文献载体标识］. 出版地：出版者，出版年，引文页码. 更新或修改日期［引用日期］. 获取和访问路径，数字对象唯一标识符.

示例：

［1］国务院学位委员会. 教育部关于加强学位与研究生教育质量保证和监督体系建设的意见［EB/OL］.（2014-03-17）［2018-02-26］. http：//old. moe. gov. cn//publicfiles/business/htmlfiles/moe/s7065 / 201403/ 165554. html.
［2］Online Computer Library Center, Inc. History of OCLC［EB/OL］.（1999-06-14）［2000-01-08］. http：//www. oclc. Org.

八、著录实例分析

例1：陈冰冰. 大学英语需求分析模型的理论构建［J］. 外语学刊，2010（2）.

例2：马聪，张建华，朱丹. 牛场称重系统的发展现状、存在问题及对策分析. 宁夏农林科技，2018，59（7）：33~35.

解析：《信息与文献　参考文献著录规则》（GB/T 7714—2015）规定：参考文献的著录内容应准确，必备项目不应缺失，以上两个例子都出现了著录项目的缺少，例1缺少了引文页码；例2缺少了文献标志码。正确的著录格式应该为：

陈冰冰. 大学英语需求分析模型的理论构建［J］. 外语学刊，2010（2）：120~123.

马聪，张建华，朱丹. 牛场称重系统的发展现状、存在问题及对策分析［J］. 宁夏农林科技，2018，59（7）：33~35.

例3：袁平华. 大学英语教学环境中依托式教学研究［C］. 北京：社会科学文献出版社，2014：37.

例4：贺哲丰，杨平. 名人典故在材料科学基础教学中的作用［G］. 中国冶金教育，2011（5）：38~40.

解析：文献类型标志是用以区分所引用的文献类型，文献类型标识码和文献类型是一一对应的，文中所引用的文献来源不同，所著录参考文献的类型也不同，由于作者不明确文献类型，常出现文献类型标识错误、混淆的情况。根据《信息与文献 参考文献著录规则》（GB/T 7714—2015）规定，普通图书标识码为M，期刊标识码为J。正确的著录格式应该为：

袁平华. 大学英语教学环境中依托式教学研究［M］. 北京：社会科学文献出版社，2014：37.

贺哲丰，杨平. 名人典故在材料科学基础教学中的作用［J］. 中国冶金教育，2011（5）：38~40.

例5：刘允芳等. 空心包体式钻孔三向应变计地应力测量的研究［J］. 岩石力学与工程学报，2001，20（4）：448~453.

例6：郭鑫年，薛彩霞，李文勤，梁锦秀，周涛. 盐碱地膜上灌及地下水补给对玉米生长发育及产量的影响［J］. 宁夏农林科技，2018，59（8）：5-8.

解析：《信息与文献 参考文献著录规则》（GB/T 7714—2015）规定：同一文献的责任者不超过3人时，全部著录；超过3人时，只著录前3名作者，中文文献其后加"，等"，英文文献其后加"，et al"。不同作者姓名之间用"，"隔开。有作者在著录的时候，会出现著者姓名遗漏的情况，如有的文章只著录第一位作者，还有的文章无论有多少作者全部著录。例5只著录了一位作者，遗漏了其他两位作者；例6则著录了全部作者。正确的著录格式应该为：

刘允芳，朱杰兵，刘元坤. 空心包体式钻孔三向应变计地应力测量的研究［J］. 岩石力学与工程学报，2001，20（4）：448~453.

郭鑫年，薛彩霞，李文勤，等. 盐碱地膜上灌及地下水补给对玉米生长发育及产量的影响［J］. 宁夏农林科技，2018，59（8）：5~8.

例7：Xiaoxia Wu, Lianzhu Zhang, Haiyan Chen. Spanning trees and recurrent configurations of a graph［J］. Appl. Math. Comput.，2017（314）：25~30.

例8：G S WANG. Resignation of Nurses in China［J］. Iran J Public Health，2014，43（1）：123.

例9：Arnold P. Lutzker. 创意产业中的知识产权［M］. 王娟，译. 北京：人民邮电出版社，2009.

解析：《信息与文献 参考文献著录规则》（GB/T 7714—2015）规定：个人著者一律采用姓在前名在后的著录形式。用汉语拼音书写的人名，姓全大写，其名可以缩写，取每个汉字拼音的首字母；欧美著者的名可用缩写字母，缩写名后省略缩写点，名字间间隔一个字符。正确的著录格式应该为：

WU X X, ZHANG L Z, CHEN H Y. Spanning trees and recurrent configurations of a graph［J］. Appl. Math. Comput.，2017（314）：25~30.

WANG G S. Resignation of Nurses in China [J]. Iran J Public Health, 2014, 43 (1): 123.

LUTZKER A P. 创意产业中的知识产权 [M]. 王娟, 译. 北京: 人民邮电出版社, 2009.

例10: 张宏, 李航, 程利冬, 等. 运用 EndNote 批量编辑加工英文参考文献 [J]. 编辑学报, 2018, (4): 369~372.

例11: 王新祥. 换热设备结垢机理的研究进展 [J]. 现代化工, 2002 (4): 22~25.

例12: 李维, 赵彬. Quantale 中的弱 S-素元及其性质 [J]. 陕西师范大学学报 (自然科学版), 2009, 37: 1~5.

例13: 黄晓庆, 孔久祥, 孔繁芳, 等. 不同寄主来源的葡萄霜霉病菌致病力测定及孢子囊大小比较 [J]. 植物保护, 2015, 3 (41): 178~182.

例14: 冯剑军, 张俊彦, 张平, 等. 基于双剪统一强度理论的厚壁圆筒塑性极限载荷分析 [J]. 固体力学学报, 2004, 2 (25): 208~212.

例15: 吴晓霞. 割点图的 Avalanche 多项式 [J]. 闽南师范大学学报 (自然科学版), 2019, 32 (02): 1~5.

例16: 张功员, 谢锡增. 编辑核对外文参考文献的效果 [J]. 编辑学报, 2006, 18 (4).

解析: 《信息与文献 参考文献著录规则》(GB/T 7714—2015) 规定: 凡是从期刊中析出的文章, 应在刊名之后注明其年、卷、期、页码。阅读型参考文献的页码著录文章的起讫页或起始页, 引文参考文献的页码著录引用信息所在页。但在实际的著录过程中, 引用期刊类文献经常会出现卷号、期号、页码的缺失或者卷号、期号标注颠倒或者页码数字标错等情况, 例10、例11缺少了引用期刊的卷号; 例12缺少了引用期刊的期号; 例13、例14混淆了卷号和期号, 出现卷 (期) 号著录倒置的情况; 例15期数为个位数, 前面不需要补 "0"; 例16缺少了引用期刊的页码。正确的著录格式应该为:

张宏, 李航, 程利冬, 等. 运用 EndNote 批量编辑加工英文参考文献 [J]. 编辑学报, 2018, 30 (4): 369~372.

王新祥. 换热设备结垢机理的研究进展 [J]. 现代化工, 2002, 22 (4): 22~25.

李维, 赵彬. Quantale 中的弱 S-素元及其性质 [J]. 陕西师范大学学报 (自然科学版), 2009, 37 (1): 1~5.

黄晓庆, 孔久祥, 孔繁芳, 等. 不同寄主来源的葡萄霜霉病菌致病力测定及孢子囊大小比较 [J]. 植物保护, 2015, 41 (3): 178~182.

冯剑军, 张俊彦, 张平, 等. 基于双剪统一强度理论的厚壁圆筒塑性极限载荷分析 [J]. 固体力学学报, 2004, 25 (2): 208~212.

吴晓霞. 割点图的 Avalanche 多项式 [J]. 闽南师范大学学报 (自然科学版), 2019, 32 (2): 1~5.

张功员, 谢锡增. 编辑核对外文参考文献的效果 [J]. 编辑学报, 2006, 18 (4): 273~274.

例17: ……EAP 教材的材料选择。最重要的原则就是真实性和难易度适当[16]

或……EAP 教材的材料选择。最重要的原则就是真实性和难易度适当[(16)]

或……EAP 教材的材料选择。最重要的原则就是真实性和难易度适当[ ]16。

解析：《信息与文献　参考文献著录规则》（GB/T 7714—2015）规定：顺序编码制要将正文中引用的文献序号置于方括号中。例 17 引文序号未放入方括号中。正确的著录格式应该为：

……EAP 教材的材料选择。最重要的原则就是真实性和难易度适当[16]。

例 18：合理使用模糊语有助于提高研究者的学术交流能力，而充分了解并正确习得模糊语策略对学术写作初学者更具意义，可帮助其在特定的学科领域中构建专业学者身份[24]。然而，基于对 22 个学术英语教材的分析，Hyland[20]指出现有教材鲜少提及模糊语的用法，甚至有极端观点认为，模糊语的使用违背了科学研究的精准原则，在学术写作中应予以避免。

解析：《信息与文献　参考文献著录规则》（GB/T 7714—2015）规定：按正文中引用的文献出现的先后顺序连续编码，将序号置于方括号中。例 18 引文顺序编码混乱，没按引文引用的先后顺序编码，较大序号首次出现在较小序号之前。

例 19[13]：关键词统计分析是文献计量学的重要组成部分，通过分析文献关键词，不仅可以研究文献内在规律，还能够揭示学术领域的学科特点、研究热点、发展方向等内容。一个关键词出现的频次越高，相关的研究成果数越多，研究内容的集中性就越强[3,4]。通过关键词的统计分析，可以揭示该学科的研究热点及研究方向。目前，关键词统计分析已经成为文献计量学研究的一种行之有效的方法[5,6,7,8,9]。

例 20：作为衡量阅读难度的标准之一，文本可读性对于阅读教学、阅读教材编排有重要意义。教学领域相关的研究者们也试图通过对教科书文本的可读性进行评估，以甄选出对于学生而言难度最适宜的语言教材[4][5]。

例 21：我国烟草种植区钾素出现明显盈余状态[2][7]。

解析：《信息与文献　参考文献著录规则》（GB/T 7714—2015）规定：同一处引用多篇文献时，应将各篇文献的序号在方括号内全部列出，各序号间用","，如遇连续序号，起讫序号间用短横线连接。例 19 在文中同一处引用不同文献，序号连续，方括号内未用短横线连接文献的起止序号；例 20 在文中同一处引用不同文献，未在同一方括号内标注；例 21 在文中同一处引用不同文献，序号不连续，未在同一方括号内全部列出，各序号间用","间隔。正确的著录格式应该为：

关键词统计分析是文献计量学的重要组成部分，通过分析文献关键词，不仅可以研究文献内在规律，还能够揭示学术领域的学科特点、研究热点、发展方向等内容。一个关键词出现的频次越高，相关的研究成果数越多，研究内容的集中性就越强[3-4]。通过关键词的统计分析，可以揭示该学科的研究热点及研究方向。目前，关键词统计分析已经成为文献计量学研究的一种行之有效的方法[5-9]。

作为衡量阅读难度的标准之一，文本可读性对于阅读教学、阅读教材编排有重要意义。教学领域相关的研究者们也试图通过对教科书文本的可读性进行评估，以甄选出对于学生而言难度最适宜的语言教材[4-5]。

我国烟草种植区钾素出现明显盈余状态[2,7]。

例 22[14]：近年来，越来越多的有关"多模态话语分析"的内容出现在外语研究中。因为单纯以语言为研究对象已经不能满足有效的话语分析，交际是运用多种符号和感官同时进行的，所以多数话语分析更加注重多模态。多模态教学模式开始受到广大学者与教师们的重视，尤其是多模态教学模式在听力教学中的应用及研究尤为广泛，而在英语写作教学应用的研究比较少，因此，多模态教学模式下的英语写作教学值得广大学者与教师们研究与探讨。

......

在大学英语教学改革发展的今天，学术英语写作为较高需求的学生提供了学习如何写本专业论文、摘要及文献的机会。多模态教学模式中多媒体资源的运用以及小组活动的分配把多模态中的符号模态与感官模态渗入到学术英语写作课堂中，使原本枯燥无味的写作课变得灵活多样，在很大程度上弥补了传统单一模态的不足，充分体现出学生为主体的课堂，提升了学生自主学习的能力。学术英语写作教学目前仍处在探索阶段，多模态教学模式的应用还有待于进一步完善，从而增强学生学术英语写作的理解力，提高其写作兴趣以及提升教师课堂写作教学效率。

参考文献：

[1] 国防. 多模态语篇图文关系识解的对比研究——以中美读者阅读英文绘本为例 [J]. 外语学刊, 2017（6）：14~18.

[2] 张德禄. 多模态话语分析理论与外语教学 [M]. 北京：高等教育出版社, 2015.

......

[10] 曾蕾, 李晶. 学术英语写作多模态教学研究与应用 [J]. 北京科技大学学报（社会科学版）, 2014（12）：27~32.

解析：《信息与文献 参考文献著录规则》（GB/T 7714—2015）规定：顺序编码制是按正文中引用的文献出现的先后顺序连续编码，参考文献表按引文的序号排列。在科技论文中，凡是引用前人已发表的文献中的观点、数据、材料等，都要对它们在文中出现的地方予以标明，并在文末列出参考文献表。例 22 在文后列出若干参考文献，但在文中并未加以标注，使参考文献形同虚设。

例 23[15]：近年来，学术语篇的互动性研究发展迅速，目前在应用语言学、心理学及语言教学等领域均取得了一席之地。据 Google Scholar 网站统计，约有 4.1 万篇学术英语文章提及了"interaction"一词，而 Web of Science 数据库中收录的互动性研究论文超过了 339 篇。Hyland 作为学术语篇互动研究的倡导者和推动者，更是发表了一系列相关文章[1-12]，并明确提出学术英语写作是一种动态的社会构建活动，在此过程中，学术作者会适当地使用多种话语策略来营造和谐的互动氛围，有效推销其观点。目前，Hyland[8] 提出的学术语篇互动模型仍是最具影响力且被广泛应用的互动理论模型。然而，现有研究多是在该模型的框架下描述相关语言资源的使用和分布特征，却较少系统地分析学术作者在使用此类语言资源时背后涉及的话语策略。因此，本文将基于学术语篇互动模型[8]，以三类最常见的立场与介入资源为例，阐释学术写作的不同互动式话语策略。

解析：《信息与文献 参考文献著录规则》（GB/T 7714—2015）规定：多次引用同一作者的同一文献时，在正文中标注首次引用的文献序号，并在序号的"[ ]"外著录引文页码。例 23 在多次引用同一作者的同一文献时，未著录引文页码。

例 24：LUTZKER A P. 王娟, 译. 创意产业中的知识产权 [M]. 北京：人民邮电出版社，2009.

例 25：霍布斯. 姚中秋，译. 哲学家与英格兰法律家的对话［M］. 上海：上海三联书店，2006：120.

解析：《信息与文献　参考文献著录规则》（GB/T 7714—2015）规定："其他责任者"著录的次序是在版本项（版本项缺省时即为出版项）之前，"译者"属于其他责任者。正确的著录格式应该为：

LUTZKER A P. 创意产业中的知识产权［M］. 王娟，译. 北京：人民邮电出版社，2009.

霍布斯. 哲学家与英格兰法律家的对话［M］. 姚中秋，译. 上海：上海三联书店，2006：120.

参考文献

［1］高鲁山，郑进保，陈浩元，等. 论科技期刊的参考文献［J］. 编辑学报，1992，4（3）：166~170.

［2］谭丙煜. 怎样撰写科学论文［M］. 沈阳：辽宁人民出版社，1982：63.

［3］冉强辉，伍烈尧，何剑秋，等. 对科技期刊学术质量评估体系中参考文献构建指标的调查和研究［J］. 编辑学报，1993，5（4）：187~192.

［4］陈浩元. 著录文后参考文献的规则及注意事项［J］. 编辑学报，2005，17（6）：413~415.

［5］马永军，倪向阳. 参考文献的评价功能及其对我国学术期刊评价的影响［J］. 编辑学报，2003，15（1）：21~22.

［6］朱大明. 参考文献的主要作用与学术论文的创新性评审［J］. 编辑学报，2004，16（2）：91~92.

［7］陈浩元. 科技书刊标准化18讲［M］. 北京：北京师范大学出版社，2000：204~205.

［8］陈志群. 科技论文文后参考文献的著录原则［J］. 科技与出版，2000（增刊）：44~45.

［9］中华人民共和国国家质量监督检验检疫总局，中国国家标准化管理委员会. 信息与文献　参考文献著录规则：GB/T 7714—2015［S］. 北京：中国标准出版社，2015.

［10］中华人民共和国国家质量监督检验检疫总局，中国国家标准化管理委员会. 中国人名汉语拼音字母拼写规则：GB/T 28039—2011［S］. 北京：中国标准出版社，2011.

［11］代君，贾强. 盾构隧道施工引起的临近地表建筑物沉降研究［J］. 山东建筑大学学报，2019，34（5）：45.

［12］张麒麟. 国外参考文献格式的比较研究［J］. 山东图书馆学刊，2019（5）：104.

［13］池营营，安珍，周小潭. 基于载文关键词统计的《编辑学报》研究热点分析［J］. 传播与版权，2019（1）：87.

［14］斯琴，全树云. 多模态教学模式下的学术英语写作研究［J］. 内蒙古财经大学学报，2019，17（5）：135~137.

［15］李晶洁，侯绘丽，宋军. 学术写作的互动式话语策略［J］. 上海理工大学学报（社会科学版），2018，40（2）：135.

参考答案

第1章 总 论

1. ①D ②A ③C ④B
2. ①showed ②judge ③Therefore ④developed ⑤assessed ⑥application ⑦conclusion ⑧research ⑨benefits ⑩summary ⑪Suggestions ⑫applying

3. A revised version for reference：

Technological advances in the 1990s enabled entirely new ways of conducting business in the US and throughout the world. Of particular importance was the World Wide Web, an enhancement of the Internet that allowed consumers and businesses to communicate in ways that were previously unavailable and perhaps even unthinkable.

From the mid-1990s to early 2000, the focus of businesses was primarily on the opportunities provided by the new Internet capabilities. Following the "doc-com" crash in early 2000, however, businesses began to recognize the problems associated with doing business on the Internet. From this point on ward, much of investors' and the media's focus shifted to companies with both Internet and "bricks and mortar" presences.

Numerous researchers have investigated on-line buying behavior over the past several years (e. g. Jarvenpaa and Todd 1997; Lohse,et al. 2001). However, little research has addressed the relationships between on-line purchasing ("clicks") and purchases made at physical stores ("bricks"). The purpose of this paper is to report the results of two exploratory studies designed to assess consumers' reactions to shopping at clicks and bricks. We first review background relevant to the two shopping channels and the challenges faced by companies operating the two shopping channels and the challenges faced by companies operating in the on-line environment. We next develop a research model and research questions to guide our investigations.

We then report the results of a large survey aimed at uncovering several of the relationships specified in the model, as well as a study of a single retailer operating in both the clicks and bricks environments. These studies provide insights into consumers' experiences with shopping on-line and at physical locations, as well as guidance to how retailers can take maximum advantage of the two shopping channels.

第2章 标题和关键词

Task 1

1. seismic composite connection; U-shaped steel beam; cyclic test.
2. hollow-core slab; precast concrete.
3. perforated steel ribs; load-carrying capacities; steel and concrete composite slabs; negative bending.
4. Web; user's personality; social media.
5. solar collector; thermal energy storage; solar thermal application.
6. cell phone; academic performance; anxiety; satisfaction; college students.
7. checkered steel-encased concrete composite beam; mechanical performance.
8. steel-concrete hybrid beams; shear connector.

Task 2

1. in; 2. in; 3. on; 4. for; 5. of; 6. for; 7. by; 8. of.

Task 3

Title: Finite element parametric analysis on bearing capacity of cast steel spherical joint

Keywords: cast steel spherical joint; finite element parametric analysis; failure mode; bearing capacity

Task 4

1. Keywords: CFST column; steel ratio; slenderness ratio; compression ratio; finite element analysis
2. Keywords: concrete-filled square steel tubular column; H-shaped steel beam; diaphragm-through connection; seismic behavior; load transfer mechanism
3. Keywords: engineering consulting services; risk assessment; project delivery methods; case studies

第 3 章 摘 要

1. With the implementation of Industry 4.0 and Made in China 2025, digital design and intelligent manufacturing become the main direction of industry development.
2. As an important part of CAD/CAM integration, Computer-Aided Tolerancing (CAT) plays a significant role in improving design efficiency and accelerating intelligent manufacturing.
3. Through the experimental analysis, it was found that the proposed image recognition and analysis method can identify the damaged buildings to a certain extent, although further research is needed to improve recognition accuracy.
4. The openness and diversity of the Internet of Things (IOT) makes it vulnerable to all kinds of malicious attacks. In order to maintain the security of IOT network space in a specific area effectively, aiming at the typical narrow band IOT (NB-IOT) in the IOT, detection and identification technology is proposed for IOT terminals based on man-in-the-middle.
5. To correct hazard cognition bias of construction workers more effectively, and prevent unsafe behavior occurring, first, hazard cognition results of the workers were obtained by psychological measurement. Then the degree of bias was reflected by contrast. At last the causes of bias were explored and analyzed. All above are based on cognitive psychology.
6. The configuration software WinCC is used to monitor, the temperature/humidity monitoring node and the device control node are connected through the gateway node. The control of the fan, valve and other equipment and environmental parameters monitoring are realized, and complete TDCS test system is built.
7. The experiment shows that the use of LORA wireless network can replace the existing traditional TDCS based on PLC. Through LORA wireless net-work, the construction and maintenance cost of the system can be greatly decreased.
8. Through experimental verification, the technology has a higher probability of detection and identification, and it has a certain application prospect.
9. This article investigated the complex permitivity and complex permeability of CNTs with the change of carbon nanotubes content.
10. With the continuous development of science and technology in China, the development of computer technology also derived artificial intelligence technology, especially under the support of big data, to the maximum extent of the advantages of artificial intelligence technology, in the computer network technology plays a great role.

第 4 章 文献综述

1. 文献综述（Review）可以是学术论文的一个组成部分，也可以是学术论文的一类，即作者在搜集大量相关文献的基础上，通过综合分析与评价，整理归纳而成的专题性学术论文。它们的写作技巧有相通之处，但是作为一篇论文，又有其独到的地方。从字面意思理解，"综"即综合，就是要求对文献资料综合分析、概括整理，使材料更加精练明确，逻辑分明；"述"即评述，要求对整理归纳后的材料进行比较专门的、系统的、全面的、深入的论述。总而言之，文献综述即对文献资料的综合和评述，是作者本人对某一方面课题的历史背景、前人工作、争论焦点、研究现状和发展前景等内容进行评论的科学性论文。

2. （1）选择主题；（2）文献搜索；（3）文献整理；（4）综合撰写。

3. 接触到一个陌生的研究领域，我们看文献的顺序应为中文综述—中文博士论文—英文综述—英文期刊文献。

中文综述是我们快速了解自己的研究领域的入口，可以帮助我们更好地认识课题。除此之外，通过中文综述可以了解该领域的基本名词、基本参量、表征方法等，这将对我们阅读英文文献时大有帮助。同时中文综述包含大量的英文参考文献，这就为我们后续的查阅文献打下了良好的基础。

中文博士论文的第1章前言或者是绪论部分所包含的信息量普遍大于一篇综述。因为它对该领域的背景以及相关理论的介绍更加详细，还会经常提到国内外在该领域拔尖的科研小组的相关研究方向。通过阅读可以更清楚地理清研究主题的脉络。

英文综述，尤其是发表在高IF期刊上或是那种invited paper，往往都是该领域的大家所写。对于此类文献要精读，分析文章的层次和构架，要特别关注作者对各方向优缺点的评价以及对缺点的改进和展望。除此之外，我们还可以从英文综述中学到很多地道的英文表达和专业名词，对我们写作能力的提升有很大的帮助。

第5章 论文主体写作

1. 论文的主体包括引言，材料与方法，结果，讨论和结论。其中引言要回答的问题是"为什么做这项研究"。材料与方法部分要回答的问题是"怎么研究"。文中应提供足够的实验细节，包含材料的来源、设备型号、实验持续时间和季节等，让其他人可遵循文中的细节进行实验复制。结果部分应回答"有什么发现"，即通过研究你的新发现是什么。讨论章节要解释研究结果意味着什么，为什么重要。与先前研究结果做对比，并解释任何矛盾。如果某些结果没达到统计显著关系，解释观察到的差异的可能，列出研究的局限性，并提出今后可进行的工作。最后，总结结论。结论部分是论文的归结部分，要写论证的结果，做到首尾一贯，同时要写对课题研究的展望，提及进一步探讨的问题或可能解决的途径等。

2. 论文引言撰写需要注意的常见问题如下：

（1）引言结构如"倒金字塔"，即从一个"宽泛的研究领域"到一个"本文要做的特定的研究"。结构不能颠倒，而且"倒金字塔"的"基底"不能无限宽广，即：不能从漫无边际的地方说起。（2）在介绍别人和自己做了什么前期工作时，要根据这一小部分的功能，有针对性、概括性地综述，不能"堆积材料"，不能照抄该文献的摘要，不要为了引文献而引文献。（3）避免使用时髦语，避免滥用套话。时髦语如"黄金催化剂的研究是当今催化研究皇冠上的钻石，是北极星"。另外，避免滥用套话的意识就是说引言部分不能到处都是"什么什么课题引起广泛兴趣，在什么中有广泛应用"。（4）指出课题组前文和本文的联系和区别。回答既然前面已经发表过几篇文章了，为什么本文还是值得发表？既然已经发过一篇快报，现在为什么要发长文章？和前文相比，有哪些新方法、新内容、新理论？是否达到更好的应用效果？如果在引言部分写清楚这些东西，审稿人的思维就会根据这些东西来判断是否可信，是否值得发表。（5）在引言部分的末尾建议用简练的话把本文的重要结果"预览"一下。但注意不能使用大篇幅预览，不能把正文里面的信息详细透露，一览无余。

实证分析，"高原鼢鼠的红细胞、血红蛋白及肌红蛋白的测定结果"这则英语导言合乎引言规范，避开了常见的问题。如 The plateau zoko rat is a kind of subterranean rat in high altitude areas, whose tunnel environment is characterized by oxygen, high CO_2 concentration, darkness, low temperature and humidity（言简意赅，明确研究领域）。Professors such as Wang Zuwang insist that this kind of environment is hazardous to ground rats, while the plateau zoko rat can reproduce in such an environment for generations, indicating that it has a special adaptation mechanism to the environment with low oxygen and high CO_2 concentration in the tunnel（别人的前期工作，针对性、概括性地综述）. In this paper, the special adaptation mechanism of the plateau zoko rats is investigated by measuring the number of red blood cells, hemoglobin concentration and myoglobin content in the blood of the plateau zoko rats（本文和前人研究的相关性，明确要研究的问题和必要性）.

3. 实验结果是对研究中所发现的重要现象的归纳，论文的讨论由此引发，对问题的判断推理由此导出，全文的一切结论由此得到。论文的结果和讨论部分写作需要注意以下几点：（1）实验或计算结果和具体的分析判断应该逐项讨论，切忌堆积实验数据和结果。（2）在描述实验结果的同时，要与以前的研究结果（自己和他人的）进行对比。（3）对背后的科学问题讨论必不可少。作者在研究中得出的见解。虽然还没有充分的证据证明或可以作为结论的，也可以进行阐述。研究与以往研究一致不一致的结果可以阐述，如果不一致一定要尽可能给出原因。（4）结语或小结，进一步提出的新问题。陈述主要发现，特别要讨论结果中的差别、研究的意义、未解答的问题及今后的研究方向讨论，用一个句子表示较为理想。

实例分析：In the existing building, additional underground space is required, and the pile supporting the upper structure is required to excavate the soil below. With the excavation of the earthwork, the stability of the pile will be reduced, and the lateral support of the pile body is an effective method to improve the stability.（研究中所发现的重要现象的归纳，引出与本文的相关性）This paper uses ANSYS to establish a two-dimensional finite element model to analyze the variation of the stability of laterally supported piles under excavation conditions.（具体分析本文中实验和设计）The results show that the pile buckling ultimate load and the supporting internal force reach the maximum when a support is added at a ratio of 0.5 times the excavation value of the pile surrounding soil.（合理使用数据，明确实验结果）The pile buckling ultimate load and the supporting internal force increase with the increase of the support stiffness. When the stiffness increases to a certain value, the increase of the load value and the internal force value is not obvious. With the increase of the support quantity, the ultimate buckling load value of the pile gradually increases, but the increase range gradually decreases.（对于实验结果的进一步讨论与分析）. Therefore, in order to strengthen the building under excavation condition, the variation of the stability of laterally supported piles should be considered, such as a ratio which makes the supports most solid and stiff.（根据实验结果，做出总结）

第6章 参考文献引注规范

1. 直接引用文献中具体内容时：

MLA：不超过三行的直接引用部分写入正文并用引号（""）标出，引号后紧跟着用圆括号夹注（Parenthetical Citation）的形式简要注明所引文献的出处，即作者姓氏+空格+页码，如是英文名，作者姓氏处只用"last name"；若是汉语名字，则只用姓氏的汉语拼音，不用汉字，页码直接用阿拉伯数字表示。句号、逗号、冒号、分号、问号等放在圆括号后，如果引用语句本身是疑问句或感叹句，问号和感叹号则要放到引号内。同样，文后要有相应的参考文献。

若所引文献作者的姓氏已在正文同一句中出现，则不需要在括号夹注中重复，夹注中只需页码即可。例如：

用方括号"[ ]"和省略号"…"标明原文变更的地方。

如果文献作者是三位以上，则夹注格式为（第一位作者的姓氏+空格+et al.+空格+页码），例如：Barker et al. 23。

四行或以上的引用则需另起一段，与上下正文之间空2行（double-spaced），左右则分别缩进10个英文字符，不加引号。若引用段落本就是一自然段，则首行需再缩进4~5个字符。在引用段落的句号之后用圆括号标明作者姓氏及页码。

APA：用引号（""）标出引用部分，引号后紧跟着用圆括号夹注（Parenthetical Citation）的形式简要注明参考文献的出处，即（作者姓氏"last name"，发表年份），必要时可加上页码，即（作者姓氏，发表年份，页码），页码用"p. +空格+阿拉伯数字"表示，所引内容若不在一页上，则页码用"pp. +空格+阿拉伯数字"表示。若引文有两个作者，其姓氏用 & 连接并标注于每次引用之后。

若 3~5 位作者，第一次引用时需列举全部的作者，往后若引用相同的文献，只需列出最主要作者的姓氏，再加上"et al."。但是，在参考文献部分，全部作者的姓名皆须列举出来。

若作者 6 位以上，则列出第一位作者即可，格式应为（作者 et al.，年份）。同样，在参考文献部分，全部作者的姓名皆须列举出来。

2. 首先，明白什么是剽窃，尤其什么是无意识剽窃。其次，端正学术态度，踏踏实实做学问。要从主观上避免抄袭行为。最后，了解引用规则，规避无意识的剽窃。

3. 该段落是按照 MLA 的引用格式。相关内容详见本章第 2 节。

第 7 章　学术论文写作中的问题汇总

1. A lot of people think…　　　不精确，"a lot"究竟是多少？
　…the weather…　　　　　　不精确，"weather" is short term
　…getting worse…　　　　　不正式
　They say…　　　　　　　　使用人称代词
　…going on…　　　　　　　不正式的动词短语
　…quite a long time　　　　不精确，时间究竟有多长？
　I think…　　　　　　　　　不正式，第一人称表达
　Research…　　　　　　　　指代模糊，谁的研究？
　…we now get…　　　　　　不正式
　…storms etc…　　　　　　指代模糊
　…all the time　　　　　　语气绝对化

2. 第一句：研究的意义；第二句：研究方法；第三句：研究结果；第四句：结论；第五句：创新点。

3. （1）引言部分。

（2）文中出现引言的地方用数字序号标注，论文最后的参考文献部分按序列出。参考文献格式为：作者名字首字母缩写．姓氏，刊物名称，卷．页码，（出版年）．

（3）paraphrase 和 summary 引用。